Learning the City

RGS-IBG Book Series

Published

Domesticating Neo-Liberalism: Spaces of Economic Practice and Social Reproduction in Post-Socialist Cities
Alison Stenning, Adrian Smith, Alena Rochovská and Dariusz Świątek

Swept Up Lives? Re-envisioning the Homeless City
Paul Cloke, Jon May and Sarah Johnsen

Aerial Life: Spaces, Mobilities, Affects
Peter Adey

Millionaire Migrants: Trans-Pacific Life Lines
David Ley

State, Science and the Skies: Governmentalities of the British Atmosphere
Mark Whitehead

Complex Locations: Women's geographical work in the UK 1850–1970
Avril Maddrell

Value Chain Struggles: Institutions and Governance in the Plantation Districts of South India
Jeff Neilson and Bill Pritchard

Queer Visibilities: Space, Identity and Interaction in Cape Town
Andrew Tucker

Arsenic Pollution: A Global Synthesis
Peter Ravenscroft, Hugh Brammer and Keith Richards

Resistance, Space and Political Identities: The Making of Counter-Global Networks
David Featherstone

Mental Health and Social Space: Towards Inclusionary Geographies?
Hester Parr

Climate and Society in Colonial Mexico: A Study in Vulnerability
Georgina H. Endfield

Geochemical Sediments and Landscapes
Edited by David J. Nash and Sue J. McLaren

Driving Spaces: A Cultural-Historical Geography of England's M1 Motorway
Peter Merriman

Badlands of the Republic: Space, Politics and Urban Policy
Mustafa Dikeç

Geomorphology of Upland Peat: Erosion, Form and Landscape Change
Martin Evans and Jeff Warburton

Spaces of Colonialism: Delhi's Urban Governmentalities
Stephen Legg

People/States/Territories
Rhys Jones

Publics and the City
Kurt Iveson

After the Three Italies: Wealth, Inequality and Industrial Change
Mick Dunford and Lidia Greco

Putting Workfare in Place
Peter Sunley, Ron Martin and Corinne Nativel

Domicile and Diaspora
Alison Blunt

Geographies and Moralities
Edited by Roger Lee and David M. Smith

Military Geographies
Rachel Woodward

A New Deal for Transport?
Edited by Iain Docherty and Jon Shaw

Geographies of British Modernity
Edited by David Gilbert, David Matless and Brian Short

Lost Geographies of Power
John Allen

Globalizing South China
Carolyn L. Cartier

Geomorphological Processes and Landscape Change: Britain in the Last 1000 Years
Edited by David L. Higgitt and E. Mark Lee

Learning the City: Knowledge and Translocal Assemblage
Colin McFarlane

Forthcoming

Globalizing Responsibility: The Political Rationalities of Ethical Consumption
Clive Barnett, Paul Cloke, Nick Clarke & Alice Malpass

Spatial Politics: Essays for Doreen Massey
Edited by David Featherstone and Joe Painter

The Improvised State: Sovereignty, Performance and Agency in Dayton Bosnia
Alex Jeffrey

In the Nature of Landscape: Cultural Geography on the Norfolk Broads
David Matless

Working Memories – Gender and Migration in Post-war Britain
Linda McDowell

Fashioning Globalization: New Zealand Design, Working Women and the 'New Economy'
Maureen Molloy and Wendy Larner

Dunes: Dynamics, Morphology and Geological History
Andrew Warren

Learning the City

Knowledge and Translocal Assemblage

Colin McFarlane

WILEY-BLACKWELL
A John Wiley & Sons, Ltd., Publication

This edition first published 2011
© Colin McFarlane 2011

Blackwell Publishing was acquired by John Wiley & Sons in February 2007. Blackwell's publishing program has been merged with Wiley's global Scientific, Technical, and Medical business to form Wiley-Blackwell.

Registered Office
John Wiley & Sons, Ltd., The Atrium, Southern Gate, Chichester, West Sussex, PO19 8SQ, UK

Editorial Offices
350 Main Street, Malden, MA 02148-5020, USA
9600 Garsington Road, Oxford, OX4 2DQ, UK
The Atrium, Southern Gate, Chichester, West Sussex, PO19 8SQ, UK

For details of our global editorial offices, for customer services, and for information about how to apply for permission to reuse the copyright material in this book please see our website at www.wiley.com/wiley-blackwell.

The right of Colin McFarlane to be identified as the author of this work has been asserted in accordance with the UK Copyright, Designs and Patents Act 1988.

All rights reserved. No part of this publication may be reproduced, stored in a retrieval system, or transmitted, in any form or by any means, electronic, mechanical, photocopying, recording or otherwise, except as permitted by the UK Copyright, Designs and Patents Act 1988, without the prior permission of the publisher.

Wiley also publishes its books in a variety of electronic formats. Some content that appears in print may not be available in electronic books.

Designations used by companies to distinguish their products are often claimed as trademarks. All brand names and product names used in this book are trade names, service marks, trademarks or registered trademarks of their respective owners. The publisher is not associated with any product or vendor mentioned in this book. This publication is designed to provide accurate and authoritative information in regard to the subject matter covered. It is sold on the understanding that the publisher is not engaged in rendering professional services. If professional advice or other expert assistance is required, the services of a competent professional should be sought.

Library of Congress Cataloging-in-Publication data is available for this book.

PB: 9781405192811
HB: 9781405192828

A catalogue record for this book is available from the British Library.

This book is published in the following electronic formats: ePDF 9781444343403; Wiley Online Library 9781444343434; ePub 9781444343410; mobi 9781444343427

Set in 10/12pt Plantin by SPi Publisher Services, Pondicherry, India
Printed in Malaysia by Ho Printing (M) Sdn Bhd

1 2011

For Rachael and Keir

Contents

Series Editors' Preface ix
Acknowledgements x

Introduction 1

1 Learning Assemblages 15
Introduction 15
Translation: Distribution, Practice and Comparison 17
Coordinating Learning 19
Dwelling and Perception 21
Assemblage Space 23
Conclusion 30

2 Assembling the Everyday: Incremental Urbanism and Tactical Learning 32
Introduction 32
Incremental Urbanism 33
Learning the Unknown City: Street Children in Mumbai 43
Learning, Rhythm, Space 47
Tactical Learning 54
Conclusion 59

3 Learning Social Movements: Tactics, Urbanism and Politics 62
Introduction 62
Knowing Social Movements 63
Global Slumming 66
The Housing Assemblage: Materializing Learning 69

	Learning and Representation: Counting the Poor	74
	Entrepreneurial Learning	85
	Conclusion	90
4	**Urban Learning Forums**	**92**
	Introduction	92
	Uncertain Forums	93
	Dialogic Urban Forums	98
	Translocalism and Translation	105
	Conclusion	113
5	**Travelling Policies, Ideological Assemblages**	**115**
	Introduction	115
	Translating Policy	117
	Comparative Learning: Translation and Colonial Urbanism	122
	Ideology and Postwar Urban Planning	128
	Neoliberal Urban Learning Assemblages	134
	Ideology and Explanation: Beyond Diffusionist Story-Making	145
	Conclusion	151
6	**A Critical Geography of Urban Learning**	**153**
	Introduction	153
	The Actual and the Possible	155
	Agency and Critical Learning	160
	Assemblage and the Critical Learning Imaginary	164
	Postcolonial Urban Learning?	167
	Conclusion	172
	Conclusion	**174**
	References	185
	Index	205

Series Editors' Preface

The RGS-IBG Book Series only publishes work of the highest international standing. Its emphasis is on distinctive new developments in human and physical geography, although it is also open to contributions from cognate disciplines whose interests overlap with those of geographers. The Series places strong emphasis on theoretically-informed and empirically-strong texts. Reflecting the vibrant and diverse theoretical and empirical agendas that characterize the contemporary discipline, contributions are expected to inform, challenge and stimulate the reader. Overall, the RGS-IBG Book Series seeks to promote scholarly publications that leave an intellectual mark and change the way readers think about particular issues, methods or theories.

For details on how to submit a proposal please visit:
www.rgsbookseries.com

Neil Coe
University of Manchester, UK

Joanna Bullard
Loughborough University, UK
RGS-IBG Book Series Editors

Acknowledgements

The idea for this book has been bubbling away for a long time, but it took me a while to figure out how I wanted the book to take shape and how to delineate its scope and contribution. There are a great many people to thank for helping with this, and the list goes well beyond the scope of these acknowledgements. The areas of my research that have made their way into the final version were gathered through various stages, and through two in particular: while a postdoctoral fellow in geography at the Open University between 2004 and 2006, which included an extended field-trip to Mumbai, and during my PhD in geography at Durham University between 2000 and 2004. All of the writing was done following my return to geography at Durham as a lecturer in 2006. I am grateful to both institutions for the support they have offered, and to all those who dedicated time and helped me during fieldwork, particularly in Mumbai, and especially activists from the Federation of Tenants Association, the Mumbai Alliance, and Slum/Shack Dwellers International.

The book manuscript itself has benefited immensely from the critical and generous readings of two anonymous readers, and the insightful comments of Kevin Ward. Kevin has been an excellent editor: always supportive and helpful in his advice, insightful, and patient. Jonathan Shapiro Anjaria also read an entire draft and offered typically incisive and supportive comments. Throughout the process of writing, I have been very fortunate to benefit from an enabling research culture of debate and experimentation at Durham. The chats, advice and support of several friends in the department have aided the production of this book immensely, and here I am particularly grateful to Ben Anderson, Angharad Closs Stephens and Gordon MacLeod. Other friends at Durham have been extremely helpful in offering advice and support during the process, especially: Ash Amin, Harriet Bulkeley, Rachel Colls, Mike Crang, Stuart Elden, Paul Harrison, Adam Holden, Kathrin Hörschelmann, Cheryl McEwan, Joe Painter, Marcus Power and Jonathan Rigg.

Outside Durham, I would like to acknowledge advice, support, helpful discussion and comments on various bits and pieces of writing in the book from a variety of people, especially Alex Jeffrey, Alex Vasudevan, Tariq Jazeel, Steve Graham, Emma Mawdsley, Steve Legg, Dave Featherstone, Mustafa Dikeç and Jenny Robinson. I would like to acknowledge the support and conversations that helped the production of the book during a fellowship in the summer of 2009 with Berlin's Irmgard Coninx Foundation, an organization with a rare commitment to genuinely international and interdisciplinary academic inquiry and debate. I am also grateful to Jacqueline Scott at Wiley-Blackwell for her gentle reminders on deadlines, and for her patience.

The biggest thank you is for Rachael, who probably found the last stages of the writing of this book more challenging than me! She has been, as ever, patient and supportive. We have had many conversations about the book's various twists and turns, often while walking through the park, from which I have benefited more than she realizes. Keir in his own unique way helped me keep the book in perspective as the writing came to a finish, and in our almost nightly conversations over the (fantastic!) 'City I Love' poems, he helped me think about how cities are learnt.

Introduction

Learning is often neglected in work on urban politics and everyday life, marginalized as background noise to urban change and the urban experience. In this book, I aim to conceptualize learning as an important political and practical domain through which the city is assembled, lived and contested, and as a critical opportunity to develop a progressive urbanism. In doing so, I address five key interrelated questions in relation to learning the city. How might learning be conceptualized? How does learning take place on an everyday basis? How does learning occur translocally? How do different environments facilitate or inhibit learning? And how might we develop a critical geography of learning? I use the concept of *assemblage* as a spatial grammar of urban learning. Assemblage is used to emphasize the labour through which knowledge, resources, materials and histories become aligned and contested: it connotes the processual, generative and practice-based nature of urban learning, as well as its unequal, contested and potentially transformative character. I develop a conceptualization of 'urban learning assemblages' in order to understand the experience and contestation of learning in different contexts: residents of informal settlements, activists working on urban informality, urban forums involving state and civil society, and urban planners and policymakers. I argue that attending to urban learning assemblages reveals important conceptual resources and empirical domains through which urbanism is produced, lived and contested.

Learning is, as Tim Ingold (2000: 155) has described it, a kind of 'wayfinding'. This is not the clichéd populist notion of learning as 'journey', but instead learning as a process in which people '"feel their way" *through* a world that is itself in motion, continually coming into being through the

Learning the City: Knowledge and Translocal Assemblage, First Edition. Colin McFarlane.
© 2011 Colin McFarlane. Published 2011 by Blackwell Publishing Ltd.

combined action of human and nonhuman agencies' (Ingold 2000: 155). Here, knowing is an uncertain, embodied process that emerges inescapably through engagement with the world around us, as Ian Borden et al. (2001: 9) put it in relation to the city: 'Knowing the city is ultimately a project of becoming, of unfolding events and struggles in time as well as in space.' If this points towards a phenomenology of urban learning, a conception of learning as an emergent property of dwelling itself, Heidegger (1971) was certainly attuned to a sense of learning as the 'plight' of dwelling, as he wrote in *Building Dwelling Thinking*: 'The real dwelling plight lies in this, that mortals ever search anew for the nature of dwelling, that *they must ever learn to dwell*.' But Borden et al.'s statement also emphasizes knowing as *struggle* in time and space, and in doing so cautions against any conception that restricts learning to localized and individual 'truth' finding. If, as Richard Sennett (2008: 289) has argued in *The Craftsmen*, 'people need to practise their relations with one another, learn the skills of anticipation and revision in order to improve these relations', in cities that practice often takes the form of conflict and struggle, and occurs in contexts of radical inequality.

Critical urbanists like Henri Lefebvre, for example, would have likely had little patience with an account of urban learning that sought no more than thick description of how people learn to interact with one another. He was more likely to ask how urban ideologies shape and limit the prospects of urban learning and, as he now so influentially put it, of a socially just 'rights to the city': 'Could urban life recover and strengthen its capacities of *integration* and *participation* of the city, which are almost entirely lost, and which cannot be stimulated either by authoritarian means or by administrative prescription, or by the intervention of specialists?' (cited in Kofman and Lebas 1996: 146; emphasis in original). There is an implicit message for critical urbanists in Lefebvre's question: that if we are interested in urban justice, then we cannot simply ask what specialist and expertise knowledge is and what it does, nor simply how learning takes place – we need alongside this to ask constantly who we learn from and with; that is, we need to attend to where critical urban knowledge comes from and how it is learnt. My focus in the book is on both the nature of learning the city and on developing a critical geography of urban learning. In doing so, one of my central concerns is with *translocal* learning: learning that is place-focused but not restricted to that place. Translocal learning involves an ongoing labour in forging and developing connections between different sources, routes and actors. My interest in translocal learning emerges in part from my desire to show how urban learning is not simply spatially bounded in local places, but is instead relationally produced, and due to the increasingly important role that translocal practices and connections play in the production of different forms of urban knowledge, notwithstanding the existence and generative influence of translocal connections in the past (Featherstone 2008; McCann and Ward 2011).

But what is learning? And how should it be differentiated from the notion of knowledge? While knowledge and learning are inextricably related processes, the focus of the book is more on learning than on knowledge. Knowledge is the sense that people make of information, which is anchored in practices, beliefs and discourses (Nonaka *et al.* 2000). This is not to subscribe to the traditional understanding of knowledge as something that people 'possess'. Rather, knowledge is located in space and time and situated in particular contexts; it is mediated through language, technology, collaboration and control; and is constructed, provisional, and constantly developing (Amin and Cohendet 2004: 30). Most importantly, if knowledge is the sense that people make of information, that 'sense' is a practice that is distributed through relations between people, objects and environment, and is not simply the property of individuals or groups alone (ibid.). In debates in organizational learning, knowledge has traditionally been separated out into the 'tacit', i.e. a pre-cognitive competence to act that is 'deeply rooted in action, procedures, routines, commitment, ideals, values and emotions' (Nonaka *et al.* 2000: 7; Polanyi 1966; Gherardi and Nicolini 2000), conceived as tied to place and difficult to move around – and the codified or explicit, i.e. that which can be 'expressed in formal and systematic language and shared in the form of data, scientific formulae, specifications, manuals and such like' (Nonaka *et al.* 2000: 7). While the tacit-codified dualism has its heuristic merit, in practice, knowledge, as a distributed social practice cannot, as Amin and Cohendet (2004: 84) put it in relation to organizational knowledge, 'be easily separated into bundles of tacit or codified knowledge or bundles of rational versus experiential knowledge'.

But what is the process of making, contesting and reproducing knowledge? This is where learning emerges. Learning is a name for the specific processes, practices and interactions through which knowledge is created, contested and transformed, and for how perception emerges and changes. The traditional conception of learning would imply a cognitive formal process of training or skill acquisition as a linear addition of knowledge, such as in learning a musical instrument or a scientific technique. In contrast, I conceive learning as a distributed assemblage of people, materials and space that is often neither formal nor simply individual. For example, people come to learn the local urban transport system, or where to buy the best fresh fruit in urban markets, or which parts of the city are safe or dangerous at particular times of day and night, not through formal training but through gradually developing a sense of how things work and change. Rather than being confined to the individual, learning as a process is distributed through relations between people–materials–environment. For example, urban transport is learnt through developing an understanding of timetables and routines, by the experience of riding on the bus or train in terms of the quality of materials or the numbers and behaviour of fellow passengers, by negotiating the contingencies of everyday interaction, and through shared stories and experiences of the nature and rhythm of transport networks. Learning here can be incremental or radical, and is as much about developing perceptions through engagement in the city as it is about creating knowledge.

Learning and Urban Change

If urban learning has been a neglected topic, it would be wrong, of course, to suggest that it has been entirely ignored in accounts of urban change. Learning has been discussed in debates on urban economies and, increasingly, in debates around travelling urban models or policies (e.g. UNDP 2003; Campbell 2008; McCann and Ward 2010, 2011; Peck and Theodore 2010). There has been a great deal of debate in economic geography, for example, about 'learning regions', 'regional innovation', 'institutional thickness', innovative 'buzz', skills development, the possibilities of knowledge mobility, and on the role of 'clusters', 'quarters', 'creative economies' and tacit knowledge (e.g. Glaeser 1999; Bunnell and Coe 2005; Florida 2002, 2005; Gertler and Wolfe 2002; Gertler 2003, 2004; Amin and Cohendet 2004; Cumbers and MacKinnon 2004, 2006; Scott 2006; Cumbers *et al.* 2008; MacKinnon 2008; Storper and Scott 2009). This work has critically engaged with, for example, the efforts of states and supranational bodies to identify and develop specialist clusters within cities and regions, often taking the form of research and development and venture capital initiatives 'which attempt to inculcate a culture of innovation and learning' and seek to 'build and reinforce a sense of cluster identity amongst constituent firms and organisations' (Cumbers and MacKinnon 2004: 959). There is a great deal of urban policy debate around, for instance, city cluster learning such as Agenda 21, or learning network formations from UN Habitat to Infocity.

As states and supranational institutions have increasingly focused on learning as central to competitive advantage within global economies, a key debate in relation to urban clusters has been around how to create linkages and networks through clustering in ways that facilitate learning through exchange and interaction. Here, proponents have advocated the importance of local links, face-to-face exchanges and 'communities of practice' (Wenger 1998; Contu and Willmott 2000; Harris and Shelswell 2005) for creating economically valuable tacit knowledge, while critics have questioned the extent to which knowledge creation is or should be 'local' rather than distanciated, and emphasize the role of external connections to and through particular organizations, the Internet, video-conferencing, exchanges across space through visits and conferences, and labour mobility (Amin and Cohendet, 2004). These debates, and their often close synergies with debates around organizational learning (e.g. Grabher 1993, 2004; Wenger

1998; Gherardi and Nicolini 2000; Nonaka *et al.* 2000; Amin and Cohendet, 2004), have informed the conception of learning put to work in the book. But as important as these debates have been, they have tended – not surprisingly given their economic focus – to restrict urban learning to questions of economic innovation, urban and regional competitiveness, and organizational learning, and have offered less in terms of critical engagement with power inequalities and exclusion, or on how learning operates as a coping mechanism or as a tactic of resistance.

The constitutive role of learning in processes of urban change and politics has been identified in debates on urban policy transfer, from Anthony Sutcliffe's (1981) *Towards the Planned City*, Ian Masser and Richard Williams's collection (1986) *Learning from Other Countries: The Cross-National Dimension in Urban Policy-Making*, and Anthony King's (e.g. 2004) surveys of colonial urbanism, to Joe Nasr and Mercedes Volait's (2003) collection *Urbanism: Imported or Exported?* and McCann and Ward's (2011) collection, *Mobile Urbanism*. Emerging work on what Eugene McCann calls 'urban policy mobilities' (e.g. McCann 2008; McCann and Ward 2010) is one important example here. This disparate work has considered, for instance, how certain cities learn from particular policy discourses, such as discourses of 'knowledge cities', 'creative cities', or neoliberal, revanchist and punitive ideologies of urban development (e.g. Florida 2002, 2005; Peck 2005, 2006; Ward 2006; Hollands 2008; Wacquant 2008; McCann and Ward 2010; Bunnell and Das 2010; Peck and Theodore 2010). If travelling urbanism is a far from new phenomenon, urbanism is none the less increasingly assembled through a variety of sites, people, objects and processes: politicians, policy professionals, consultants, activists, publications and reports, the media, websites, blogs, contacts, conferences, peer exchanges, and so on. But despite this surge in critical literature on travelling urban knowledge and policy learning, there has been little attempt to consider how learning itself might be conceptualized.

More broadly, debates about the role of learning have a long history, particularly in relation to economic development, whether Schumpeter's (1934) *Theory of Economic Development*, Hayek's (1945) paper *The use of knowledge in society*, Machlup's (1962) *Production and Distribution of Knowledge*, or contemporary debates on the 'knowledge economy' or the World Bank-led 'knowledge for development' initiatives (e.g. Leadbeater 2000; King and McGrath 2004; Leydesdorff 2006; McFarlane 2006a; Johnson and Wilson 2009). There is also, of course, a similarly complex history of debates on the nature, production, meaning and value of learning, with disparate influences from theories of epistemology (Polanyi 1966, 1969; Greco and Sosa 1999), genealogies of historical knowledge (Foucault 1980), science studies (Latour 1999; Callon *et al.* 2009), organizational studies (Wenger, 1998; Amin and Cohendet 2004), cognitive anthropology (Hutchins 1995; Ingold 2000), and critical education and pedagogy

(Freire 1970; Fejes and Nicoll 2008). If these debates are varied and distinct, all of them contain one central claim or assumption about learning: that learning is a process of potential transformation. Learning, even where it is explicitly described as uncertain – as in, for instance, strands of organizational theory that emphasize creativity and invention – refers to a process involving particular constituencies and discursive constructions, entails a range of inclusions and exclusions of people and epistemologies, and produces a means of going on through a set of guidelines, tactics or opportunities. As a process and outcome, learning is actively involved in changing or bringing into being particular assemblages of people–sources–knowledges. It is more than just a set of mundane practical questions, but is central to political strategies that seek to consolidate, challenge, alter and name new worlds.

Locating Urban Learning

My approach in the book is to offer a conception of learning and then to bring that conception to urbanism. My starting point is not to attempt to identify what it is about urbanism that involves learning; instead, my approach has been to begin with a conception of what learning is before we can locate it in the city. There are three rationales for this approach. Firstly, and pragmatically, it reflects my own intellectual biography – I first became interested in conceptualizing learning, and then interested in how that conception of learning might illuminate our understanding of urbanism. Secondly, the main alternative to my approach of identifying urban learning by working from a conceptual toolkit of what learning itself involves, would have been to attempt to identify urban learning by working from particular encounters people have with urban forms and processes. While this approach would have been entirely plausible, there is a risk here of operating under the pretence that we can pinpoint how we learn the city through specific urban causal moments, that is, to reduce the ways in which urbanism is learnt to urban encounters, forms and processes. As different groups of people – urban residents, activists or policy-makers for instance – learn cities, that learning is not *only* a product of the encounter with the city itself. It is not simply a product of, say, specific encounters with urban housing materials, or of street layouts and architecture, or of policy domains defined by, say, urban statistics on healthcare in different parts of the city. Rather, the learning that emerges is a combination of two broad constituent parts: the changing nature of urbanism itself, and individual and group experiences, perceptions, concerns, interests, agendas, memories, hopes, fears, and so on. Urban learning is no more dictated purely by the city itself any more than it is dictated purely by individual experiences and perceptions. Learning emerges through a relational co-constitution of city and individual, where the individual's experiences, perceptions, memories, agendas and ways of inhabiting the city cannot be read as *urban* experiences alone.

Another way of putting this is to say that urban learning is not exhausted by the specificity of particular encounters with urban form or process, but is instead embedded in the current of people's lifeworlds and is shaped relationally. I am not suggesting that it would be wrong to take the approach of identifying urban learning from specific urban forms, processes and encounters in the lives of particular people, and I am sure that it is possible to embark on such a project without losing sight of the relational nature of learning, but given the risks in pursuing this, my approach has been to develop a theory of learning itself and to then consider how it takes place and functions in various urban domains. Following this, the third reason for beginning with a general theory of learning and then using it to understand particular forms of urbanism is that the learning of cities does not have any pre-given geography. That is, urban learning does not necessarily take place 'at large' in the city, for example in the street or neighbourhood or public space – although as I will argue, it does take place in all of these places – but potentially in any number of places. For example, it may be that a key transformatory moment in how a particular person learns a city, urban form or process occurs not in the spaces of the city itself but through a conversation with a friend on a car journey, or on a rural holiday, or through a website or reading a book, or in a conference venue. Urban policy-makers, for instance, might learn the city in large part through databases and spreadsheets, past policy documents or conversations with policy-makers at conferences well away from the city they work in. Given that we cannot know the particular sites and sources of urban learning, given that we cannot identify the geography of urban learning in advance and go out and research it, my approach has been first to understand learning itself, and then to consider how it emerges and operates through particular urbanisms – for example, within informal settlements or in the work of urban policy-making – whilst recognizing that these are only particular sociospatial instances of urban learning for the individuals and groups in question.

I begin, then, with a theory of learning itself, and then consider how that theory operates in relational constitution with and through the city. This means, to take two examples, that while the nature of the 'tactics' I discuss in the book (Chapters 2 and 3), or the form of the issues explored through 'forums' (Chapter 4), may be specific to urban spaces, processes and concerns, it is equally plausible that the conception of learning developed here might be applicable to non-urban contexts. If specific forms of learning discussed in the book are particular to urbanism – for example, to the sociomateriality of informal settlements – the *theory of learning* that I develop could potentially be used to illuminate not just urban production, contestation and possibility, but other domains outside of urbanism (such as the state or the village). That said, cities – as spaces of encounter and rapid change, of concentrations of political, economic and cultural resources, and of often confusing illegibility – demand to be learnt and learnt again by different people and often for very different reasons, from coping mechanisms and personal advancement to policy-making and questions of contestation and justice.

In what sense does the city *demand* learning? What distinguishes cities is their density, changeability and complexity. They can be places of anchoring, security, ease and sociality, as much as equality, opportunity, struggle, conflict, exploitation and hardship. They are characterized increasingly by profound density – for instance in growing informal settlements, where one in three urbanites now live (Davis 2006) – as well as spatial extension in the form of sprawl or corridors of communication and economy. They can be extraordinarily diverse in their spatial experience, from city centres that transform between day and night, to patchworks of industrial and service areas, shopping centres, abandoned spaces, parks, and often radically varied architecture. They change with rhythms of commuters and tourists, schools and nightlife, and are often the locus of new politics, lifestyles, subcultures, imaginaries, and technologies. They are constituted, as David Harvey (1997: 229) has put it, by 'conflictual heterogeneous processes which are producing spatio-temporalities as well as producing things, structures and permanencies in ways which constrain the nature of the social process'. The diverse and changing nature of the urban experience, the transformatory nature of many cities themselves, and the changing concerns, agendas and lives of those of us who live urban lives, means that our perception and knowledge of the city, whether in the shape of our city of residence or in relation to urbanism more generally, is altered over time, and that we inevitably learn and relearn the city, sometimes incrementally and sometimes radically.

This is not to say that people are constantly and actively attempting to formally learn the city in an explicit way – although sometimes this is what happens, for instance as we attempt to understand a new neighbourhood that we might be moving to or as we move through cities with a map, underground timetable or architectural guidebook – as if a fear of uncertainty drives people into a desire to have to know how the city is changing. This would be to misunderstand learning as a purely cognitive process of acquiring information in a linear way, when in practice it is often an implicit experiential process of incrementally changing how we inhabit or perceive urban space that occurs without our realizing it. Nor is it to say that everyone learns the city in the same kind of way or to the same sort of extent. But whether you are a resident, a visitor, a policy-maker or an activist, the changing nature of the city, of how it is experienced over time and space, and of the particular contexts, agendas and interests of different individuals and groups, inevitably alters knowledge and perception of it. This book is about some of the ways in which that takes place.

The book consists of six substantive chapters followed by a conclusion. In Chapter 1, I develop a conception of 'urban learning assemblages' that serves the rest of the book. Firstly, I conceptualize learning based on three closely interrelated processes: *translation, coordination* and *dwelling*. Translation refers to the relational and comparative distributions through which learning is produced as a sociomaterial epistemology of displacement and change; coordination refers to the construction of functional systems that enable learning as a means of coping with complexity, facilitating adaptation and organizing different domains of knowledge; and dwelling refers to the education of attention through which learning operates as a way of seeing and inhabiting the world. Each step in the argument focuses on the importance of appreciating learning as a distributed process that foregrounds materiality and spatial relationality. Secondly, I use *assemblage* to develop a spatial grammar of learning focused on relations of history and potential, on the making of spaces of learning through practice, and to highlight how spaces of learning are structured through unequal relations of knowledge, power and resource.

In Chapter 2, I examine how the city is learnt on an everyday basis, and consider practices as different as housing construction and maintenance within informal settlements and the lives of street children in Mumbai, to skateboarding, different forms of walking, and making a living. I offer two interrelated conceptualizations of how learning features in the production and contestation of everyday urban life: *incremental urbanism* and *tactical learning*. Incremental learning, as laborious and historical accretion of knowledge rather than linear addition, is central to learning through dwelling, and is common to a whole range of urban processes and forms. For example, everyday learning might involve shifts in the perception of what urban materials do as they are assembled through different relations and interactions. In this sense, 'improvisation' emerges not necessarily as a sudden change, but as a creative recasting of relations that results from everyday dwelling. If incremental learning connotes a gradual temporality of urban change, the chapter goes on to reflect upon the multiple spatiotemporal rhythms – from walking to longer-term temporalities like migration – that are negotiated in part through urban learning. The final part of the chapter focuses on tactical learning. While tactical learning can emerge from modes of urban dwelling such as incrementalism, it can also emerge through forms of resistance that stand apart from everyday dwelling. As we shall see in the example of the Federations of Tenants Association, based in Mumbai, tactical learning can originate in learning through coordination and translation, and can radically open the possibilities of urban dwelling. The chapter is based on fieldwork conducted in Mumbai between 2005 and 2006 and in São Paulo in 2008 on the production and maintenance of informal settlements.

In Chapter 3, I argue that in contrast to a tendency in debates on social movements to background processes of learning, learning is central to how activists in urban social movements develop forms of organization and political strategy. I examine the politics of urban learning in the work of Slum/Shack Dwellers International (SDI), especially in relation to its Indian chapter, known as the Alliance, and argue, firstly, that central to SDI's work is a sociomaterial practice of learning in groups; secondly, that the nature of urban learning in SDI is key to the formation of political organization; and, thirdly, that this urban learning is critical to the construction of a particular political subjectivity of social change. I describe how urban learning assemblages are formed and function in SDI through the example of tactical learning in model house construction. I go on to examine the learning of political organization in SDI through, in particular, its tactic of enumeration, or self-census, of informal settlements. The chapter uses the notion of assemblage as an orientation towards the agency of materials – including self-census maps, charts, enumeration documents and model house materials – in the learning practices of SDI's work, and shows the importance of documentary representations within urban learning assemblages as they travel through and give shape to multiple sites and practices. I end the chapter with a critical discussion of how SDI's learning practices around house construction and enumeration relate to and construct particular political subjectivities and conceptions of social change. If Chapter 2 focuses more on dwelling than on the translation and coordination elements of urban learning assemblages, this chapter demonstrates the importance of translation and coordination.

The discussion of SDI draws on fieldwork conducted on SDI, especially on the Alliance, during several research visits to Mumbai, particularly two trips between October 2001 and March 2002 (during my PhD research), and November 2005 and June 2006. The data used in the chapter emerges from repeated interviews and meetings with over 30 members of the Alliance and other members of the SDI movement either based in or passing through cities in India, the UK, Brazil and Germany, as well as analysis of grey literature and observations of their work. SDI's incredibly voluminous amounts of literature and online resources must, of course, be read with a degree of caution. Like many social movements, SDI's writings are very much part of its political project of producing an image of itself as innovative, skilled, unique, and devoid of unequal relations of power, resource and knowledge, and as such should be approached with a critical eye.

In Chapter 4, I examine the sorts of environments that might give rise to progressive forms of urban learning by developing the idea of *urban learning forums*. The forum is a particular type of urban learning assemblage in that it signals the production of a centralized and organized environment specifically geared towards learning between different actors, including the state, donors, non-governmental organizations, local groups, researchers and activists. The forum is an example of learning through coordination in that it centralizes and translates multiple different forms of knowledge from different people and groups. I pick up on some high-profile examples of learning in participatory urban forums such as those in Brazil, and examine the role of different actors in these processes. The chapter argues that the success of participatory urban learning forums depends upon: firstly, and drawing on Callon *et al.*'s (2009) study of science and technology controversies, the *intensity, openness* and *quality* of these forums; secondly, the commitment of state authorities to the participation of the marginalized and the poor and to ceding decision-making powers to the forum; and thirdly, pressure from civil society. In the second part of the chapter, I consider whether urban learning forums that take place in the context of *translocal* urban learning assemblages across the global North–South divide face particular challenges that accompany that very global categorization. I develop this argument through the example of an exchange between the Indian Alliance previously discussed in Chapter 3, and a London-based homeless movement called Groundswell. As this and other examples show, urban learning forums that cut across the global North–South divide raise the spectre of this form of categorization. Categories like North and South, First and Third Worlds, developed and developing, can function to militate against the prospect of translocal urban learning forums.

While the obstacle to the success of translocal urban learning in this example is a particular kind of perception of the global North–South divide, the more general point at stake here is that the success of translocal urban learning experiments depends upon a commitment to learning through translation – i.e. through difference rather than in spite of it – rather than simply through an attempt to learn through cities which appear to be 'similar'. The research for this discussion was conducted between 2001 and 2002 in the UK, both through interviewing activists at Groundswell in its networks and attending its exchanges, and through SDI meetings in the UK, especially those organized through the SDI-donor Homeless International based in Coventry, and follow-up interviews with SDI activists in Mumbai and with SDI supporters in London.

Chapter 5 attempts to build upon a recent upsurge of interest in travelling forms of urban policy and planning by considering how ideology shapes the nature of policy and planning learning. I aim to contribute to this literature in two main ways. Firstly, by offering a critical framework for conceptualizing urban policy mobilities, I highlight four key issues: firstly, the *power* at work in policy learning, i.e. the forms of power that promote, frame or structure particular kinds of learning; secondly, the *object* of learning, i.e. the epistemic problem-spaces that the mobility of policy and planning creates and addresses; thirdly, the *form* of learning, i.e. the organizational nature of learning; and fourthly, the *imaginary* at work in learning, i.e. the image of urban reassembling that learning seeks to accomplish. The aim of this framework is to examine critically the ideologies and inequities of mobile policy learning, as well as the role of the specific agents that constitute these urban learning assemblages.

Secondly, in developing this framework, I aim to unsettle the presentism of urban policy mobility scholarship. If the extent and speed of policy mobility has increased – driven by the growing amount of travel, conferences, study tours, policy networks, and use of the Internet – urban policy mobility is, of course, far from new. Planners, architects, policy-makers and consultants have *always* sought to learn from elsewhere in their attempts to assemble the city, so much so that the city is already a relational product of different agendas and strategies from other cities. In order to demonstrate different contexts, logics, and forms of mobile urbanism, I examine quite specific urban learning assemblages from three particular periods.

Firstly, I discuss learning as translation in the example of colonial urban learning. Focusing on urban planning in colonial Bombay (now Mumbai), I argue that comparison – as a central form of learning as translation (Chapter 1) – was central to how urban planners learnt the city, by exploring how urban planners formulated solutions for Bombay's sanitation crisis through comparative learning with British cities. Secondly, I discuss mobile urban planning initiatives in the 1950s driven by Cold War ideologies of modernism and socialist realism, taking the example of the influential architect Constantinos Doxiadis and his plans for Baghdad, as well as deeply ideological planning initiatives for East Berlin during the German Democratic Republic (GDR). Doxiadis and his visual plans became a central actor in a translocal learning assemblage that combined scientific modernist planning with a capitalist vision of the city. In contrast, socialist realism explicitly opposed modernist planning, and in East Berlin sought to propagate a socialist vision of the city that combined local architectural traditions with Soviet monumentalism. As with colonial Bombay, these are sociomaterial conceptions of the city that sought to reassemble urbanism through an ideologically structured form of learning.

Thirdly, I consider the nature and implications of neoliberal ideology both for the kinds of urban policy that become mobile, and for ways in which learning is framed, translated and contested. I draw upon a variety of examples here that illustrate how ideologies of urban 'development', urban 'recovery', and of the 'creative' or 'smart' city, structure contemporary urban policy learning assemblages. Particular agents become important in framing these neoliberal forms of mobile urban policy, including institutions like the World Bank or the Manhattan Institute, or influential individual consultants like Richard Florida and William Bratton. In the three different periods that the chapter examines, I reveal how ideology can shape the nature of urban learning, and the very range of urbanisms examined is an attempt to show that the critical framework of reassembling policy urbanism that the chapter deploys is one that can be usefully deployed to grasp how learning from elsewhere plays an important role in how cities are reassembled. The research for this chapter took the form of archival research in Mumbai in 2006, during which I collected material on the development of sanitation debates and infrastructure in mid-nineteenth century Bombay as part of a larger research project on sanitation in Mumbai. This was a particularly important period for the roll out of colonial sanitation infrastructures and debates, and I drew principally upon municipal corporation reports and newspaper coverage from the time.

Chapter 6 develops a critical geography of urban learning. It does so by first outlining a critical praxis of *evaluating* existing dominant forms of urban knowledge and learning (incorporating the framework developed in Chapter 5), *democratizing* the sorts of knowledges and groups that constitute how dominant urban knowledges are learnt (incorporating the discussion of urban learning forums in Chapter 4), and *proposing* alternative, more socially just forms of urban learning through this democratized learning. The chapter then seeks to deepen and extend this schema of evaluate–democratize–propose by, firstly, examining what the concept of assemblage offers critical urban learning and, secondly, considering how a critical geography of urban learning might respond to the translocal challenge of a postcolonial urban learning project. Assemblage offers three orientations to a critical geography of urban learning: firstly, a descriptive focus on learning as produced through *relations of history and potential*, or the actual and the possible; secondly, a reconsideration of the *agency* of learning, particularly in relation to distribution and critique; and thirdly, a particular critical *imaginary* through the register of cosmopolitan composition. Responding to the postcolonial challenge demands careful consideration of who is involved in the production of critical geographies of urban learning – the 'democratize' imperative of the evaluate–democratize–propose strategy. Here, I consider the important role of comparison in translocal urban learning first discussed in Chapter 1 and then in Chapters 4 and 5, and argue for a conception of comparison as a mode of thought produced through an ethico-politics of multiple theory cultures.

In examining these different forms and contexts of urban learning in the book, I am not suggesting that there are forms of learning that are somehow more 'authentic' or 'real'. I do not argue, for instance, that the everyday learning examined in Chapter 2, or the learning of activists in SDI in Chapter 3, is more of an accurate reflection of urban reality than that of policy-makers (Chapter 5) or urban researchers (Chapter 6). Equally, I am not suggesting at all that SDI, for instance, offers a model or paradigmatic example of urban learning that researchers or indeed cities should privilege or adopt. For a start, SDI's particular forms of learning, as we will see, are not without their critics, remarkable as those learning experiments may be. What I do argue for is a democratization of urban learning, and it is true to say that SDI provides elements of this in its commitment to placing the knowledge of the marginalized at the centre of urban development. As I argue in Chapter 4 on urban learning forums, there is a radical potential of socially just transformation inherent in the possibilities of urban participation, as long as that participation is meaningful, sustained, and has the capacity to make decisions, for instance around urban redistribution. This is why the example of participatory budgeting from Porto Alegre discussed in that chapter is of such crucial importance as a demonstration of what can be achieved. In Chapter 6, I expand on the sort of urban learning I advocate by outlining a critical geography of urban learning. It is this *process* of a critical geography of urban learning assemblages, and not a particular body of content or indeed a particular model, that I argue for through the book.

If Le Corbusier's (1923) infamous injunction was that 'a house is a machine for living in', the provocation in this book is that *the city is an assemblage for learning*. The Swiss architect's machine aesthetic, rhapsodically expressed in *Towards a New Architecture*, was to no small extent a thesis of unlearning architecture and relearning urbanism. Gone, he insisted, were the craft skills of carpentry, masonry and joinery; instead houses should be built 'all of a piece', made by machine tools in factories and assembled through Fordist production lines. With one cursory eye to history and the other firmly on the possibilities of mass production, he argued that new rules and standards of assembly had to be learnt in order to mass produce housing through principles of geometric and functionalist efficiency, housing as a construction form with more in common with lightweight car bodies than the material diversity of the past. Whatever the strengths and shortcomings of Le Corbusier's claims and designs, the implicit but nonetheless crucial invocation of urban learning and relearning remains largely ignored in urban studies. My hope is that this book goes some way to addressing that.

Chapter One

Learning Assemblages

Introduction

This chapter offers a conceptualization of learning. My aim is to consider what learning is, and to begin to think through what that might mean for thinking and writing about urbanism. The theory of learning that I develop is intended to then take shape relationally through the urbanisms discussed in the subsequent chapters. At its most general, learning involves either the acquisition of knowledge or skill, and/or a shift in perception from one way of seeing a problem, issue, relation or place, to another. It is not necessarily explicitly cognitive. Skills can be implicitly acquired, for example, through the experiential practice of craft. Learning embodies a transformation of knowledge, and/or perception, and/or self, and can be a process of control and ordering, or confusion and instability. It can arise from repetition, from performances on the wing, structured training, autonomous experimentation, events that interrupt the 'known' or that lead to a new way of seeing, and more. Rather than presupposing, as Tim Ingold (2000: 416) has put it, that 'a body of context-free, propositional knowledge – namely a technology, or more generally a culture – actually *exists* as such and is available for transmission by teaching', learning emerges through practical engagement in the world (see Lave 1988).

Learning is distributed as, in Callon *et al.*'s (2009: 58) description, 'embodied forms of know-how, knacks, knowledge crystallized in various materials, and craft skills', and is often an uncertain affair, for instance in relation to moments of creativity and invention. As a practice-based distribution, learning involves particular constituencies and discursive constructions, entails a range of inclusions and exclusions of people and

Learning the City: Knowledge and Translocal Assemblage, First Edition. Colin McFarlane.
© 2011 Colin McFarlane. Published 2011 by Blackwell Publishing Ltd.

epistemologies, and produces a means of going on through a set of guidelines, tactics or opportunities. As a process and outcome, learning is actively involved in changing or bringing into being particular assemblages of people–sources–knowledges. It is more than just a set of mundane practical questions; it is central to the emergence, consolidation, contestation, and potential of urban worlds. In this chapter, I first offer a conceptualization of what learning is in practice, and, second, consider how we might conceptualize the spatialities of learning through the notion of 'assemblage'. In doing so, I develop a conception of learning that serves the rest of the book by making three arguments in relation to learning.

Firstly, learning is always a process of translation. This underlines the importance of intermediaries in the production of travelling knowledge; the spaces and actors through which knowledge moves are not simply a supplement to learning, but are constitutive of it. Secondly, and following on from this, learning is not simply a process of translating knowledge through space or accessing stored data, but depends on the (re)construction of functional systems that coordinate different domains. Thirdly, while learning can be structured through the inculcation of facts, rules, ideas or policy models, in substantive practice learning operates as the 'education of attention' (Gibson 1979; Ingold 2000). This means that learning can entail shifts in ways of seeing, where 'ways of seeing' is defined not simply as an optical visuality, but as haptic immersion. These three interrelated aspects can be summarized as *translation, coordination* and *dwelling*. Each step in the argument focuses on the importance of appreciating learning as a distributed process that foregrounds the materiality and spatial relationality of learning. In addition, in each of these three areas there is an important set of ethico-political concerns around how learning occurs, what sorts of urbanisms are privileged, and the potential role of various constituencies within that, including activists, policy-makers and researchers. In order to advance this argument, I will draw on a wide terrain of debates that have approached learning, including within geography, organization theory, science studies, cognitive anthropology, postcolonial studies, and urban studies.

I bring this conception of learning to a particular conception of assemblage. While the concepts of translation, coordination and dwelling are thought of as spatial processes in the chapter, they do not in themselves provide a theory of the spatialities of learning. It is in this context that I use assemblage to highlight how learning is constituted more through sociospatial *interactions* than through the properties and knowledges of pre-given actors themselves, and to think of the spatialities of learning as relational processes of composition. As I will explain, I use assemblage both as a concept and as an orientation by emphasizing three important spatialities of learning. Firstly, assemblage locates the constitution of learning in *relations of history and potential*, or the actual and the possible. Assemblage draws attention to the particular alignments produced through multiple spatiotemporalities of translation, coordination and dwelling, and to how they are reconstituted through different relations and contexts. Secondly, and following this, assemblage signals how learning is produced not simply as a spatial category, output or resultant formation, but *through doing, performance and events*. Thirdly, and finally, assemblage emphasizes how learning is *sociospatially structured, hierarchalized and narratavized* through unequal relations of knowledge, power and resource. While I illustrate many of the arguments in this chapter through urbanism, I concentrate the discussion on developing a conception of learning assemblage that will be applied to urbanism in the subsequent chapters.

Translation: Distribution, Practice and Comparison

Translation offers four perspectives to a conception of the constitution of urban learning through the creation and transformation of knowledge: a focus on distributions; a concern with intermediaries and displacement; as partial, multiple and practice-based; and as produced through comparison. Firstly, translation challenges the diffusion model that traces the movement of knowledge as innovation (Latour 1986, 1999). While the diffusion model focuses on travelling knowledge as the product of the action of an authoritative centre transmitting knowledge, translation focuses on travel as the product of what different actors do in and through *distributions* with spaces and objects, from artefacts and ideas to products and models (Gherardi and Nicolini 2000: 335). That is, translation emphasizes the materialities and spatialities through which knowledge moves and seeks to unpack how they make a difference to learning, whether through hindering, facilitating, amplifying, distorting, contesting, or radically repackaging knowledge. For example, as Chapters 3 and 5 will show, urban activists and policy-makers learn in part by translating knowledge through models and documents that move through multiple spaces, from resource centres and conference meeting rooms to Internet sites and chats over coffee. This serves to remind us that urban learning through translation is not reducible to urbanism *per se*, but to a diverse host of encounters across multiple space-times.

Secondly, and crucially, this draws attention to the importance of various forms of *intermediaries*, and promotes two inseparable relational perspectives: the importance of relations between the 'near' and 'far' in producing knowledge, for instance in the ways in which the Internet or an exchange of activists or policy-makers may make distant actors proximate; and the agentic capacities of materials in producing learning, for example, the differential and contingent role of urban plans, documents, maps, databases or models in producing, shaping and contesting urban learning (Amin and

Cohendet 2004). These intermediaries matter; translation is open to the possibility of varying degrees of stability and flux. It is not the case that every encounter must always involve change, nor is it the case that every encounter must always involve the recreation of a periphery in the image of a centre. Consequently, translation positions learning as a constitutive act of world-making, rather than occurring prior to or following from engagement with the world. It positions learning as, to paraphrase Derek Gregory (2000) writing in the context of colonial cultures of travel, an *epistemology of displacement* in which travel is not a mere supplement to learning, but constitutive of it.

Thirdly, given the focus on intermediaries and distributions, the geographies of translation centre on the idea of practice. The attention to practice collapses traditional dichotomies that separate, for example, knowing from acting, mental from manual, and abstract from concrete, that continue to contour ontologies of learning (Polanyi 1969; Hutchins 1995; Wenger 1998: 48). If we reject the functionalist view of knowledge as static, bound and fixed, and argue instead for a view of knowledge as social, then the practices and materialities through which knowledge is learnt are brought into view. Learning is a process of heterogeneous engineering that demands a relational materialism; for instance, a range of materials, from commodities and shops to public art, parks and infrastructure, make a difference in the production and movement of urban knowledge (Thrift 2007; Graham and Thrift 2007; McFarlane 2009b). The attention to practice reveals the partial and multiple nature of learning. Learning is territorialized through various forms of inclusion and exclusion, meaning that it can be to varying intensities in or out of the 'proper' spaces (Law 2000). The notion of 'situated knowledge', popularized most notably by Donna Haraway (1991), underlines partiality by focusing on the embodied nature and contingencies of knowledge production. The emphasis on the situatedness of knowledge also reminds us that practice is not simply of the present, i.e. of the immediate encounter in the city, but can also be a practice of individual or collective remembering or imagination oriented towards the past or future. But while situated, this knowledge is also mobile: it is formed not simply in place but through multiple knowledges that run through and call into being various spaces.

Fourthly, and finally, a key form of learning through translation is *comparative* learning. Urbanism, for example, has always been conceived and known comparatively. I am referring here not just to explicit forms of comparison – comparing city A with city B, for instance – but implicit comparisons that to different extents constitute how claims are made about the city. When we read a study of a particular city, we often find ourselves comparing the arguments, claims and instances with other cities that we ourselves study or know of. The implication is that claims about 'the city', or about a particular form of urbanism, are an implicitly comparative claim, because our claims and arguments are always set against other kinds of urban places, experiences, possibilities or imaginaries. And yet we rarely acknowledge this in urban studies. Here, I am thinking of comparison not as an explicit research methodology, but as an implicit mode of thought that informs how we construct knowledge and theory of the urban – in short, comparison as a crucial site in how we learn about what the urban *is*. Comparison is not just a spatial register of learning cities; there is also a temporal dimension as we learn in relation to our memories of past experiences and cities. Taken together, these four elements of translation – distribution, intermediaries, practice and comparison – identify sites and methods through which learning functions by creating or changing knowledge or perception.

Coordinating Learning

Translation always occurs in relation to multiple sites and objects, meaning that it requires coordination. Anthropologist Edwin Hutchins (1995) showed how distributed knowledge shifts learning from individual decisions or actions to allocations of collective agency, and indeed enables the agency of that collective. This requires, in Hutchins' terms, seeing learning as 'softening' the boundary between individual and environment: 'Learning is adaptive reorganization in a complex system' (Hutchins 1995: 288, 289). In these distributions, different phenomena act as organizing devices in learning, what Hutchins called 'mediating structures'. In cities, these devices might be as varied as language, models, procedures, rules, documents, instruments, traffic lights, market layouts, ideas, discourses, and so on (ibid. 290). They are not, however, necessarily forms of codified knowledge: individuals, such as the leader of an urban social movement, can coordinate different forms of tacit and codified knowledge in the communication of new ideas or strategies to members of the movement (Chapter 2). One example Hutchins used was that of the written artefact. In order to put a written procedure to work, people must coordinate with both the procedure and the environment in which the actions are to be taken. Words, meaning, document and world coordinate with each other over time, producing a kind of 'situated seeing' that makes it difficult to clearly demarcate the individual and the outside (ibid. 300), meaning that it can be difficult to locate learning as 'belonging' to one or the other.

Hutchins' discussion of the written artefact reminds us of the performative role of representation within learning, and insists that learning depends upon constantly *constructing functional systems that coordinate different domains*. Coordination is a process of sociomaterial adaptation. Fischer (2001), for example, showed how an urban planning experiment bridged a range of different interests across space through the assistance of an interactive electronic table – acting as a coordinating tool to align different actors by enabling people to jointly design and edit an urban layout (Amin and Roberts 2008a: 362). Sennett (2008: 127–9) discussed learning as coordination in relation to 'domain shifts', referring to a practice or form being translated through multiple sites. For example, urban plans (e.g. of infrastructure) coordinate domains as different as science, engineering, and social policy by instigating a chain of translation between them. Sennett (2008: 201–5) provided an example of domain shift in relation to the seventeenth-century polymath Christopher Wren, who was tasked with designing a plan for London following the Great Fire of 1666. Wren drew upon his experience as a scientist in his urban plans; he drew upon the principle of circulation of blood to imagine streets as arteries and veins of free-flowing traffic, goods and people (see Joyce 2003); he drew upon telescopic imagery to imagine streets as disappearing into the distance of deep space; and he drew upon microscopic imagery to plan for urban density, for instance in the number of churches for people in different parts of the city. Wren was using the medium of the plan to coordinate domains as different as perspectives from science, including the particular ways of seeing that scientific instruments afford, the body, mobility, infrastructure, services, and the hope that a devastated London could be planned anew. These domain shifts – a kind of 'reformatting' (Sennett 2008: 210), or 'learning-by-switching' (Grabher and Ibhert 2006: 261) – constitute coordinating devices that involve relays of translation, and that can stimulate the imagination in learning new kinds of urbanism.

The list of urban coordination tools is, then, a long one, and includes sites as mundane as travel timetables or maps as well as policy documents, urban census databases, statistical databases of urban labour markets and investment histories, one-off events like policy conferences, study tours, exchanges of activists, and the town-hall meeting. They can function as what Latour (1999) has called 'centres of calculation' in that they combine different forms of knowledge to make calculation possible. Coordination devices are not, of course, neutral: there is often a politics to how they operate and are constituted, especially in relation to how different forms of urban knowledge are coordinated, and in the potential of that coordination to facilitate more socially just – or indeed socially unjust – forms of urbanism. As Chapter 4 will argue, a key coordinating device in this respect is the *urban forum* – a particular type of centralized urban learning environment, explicitly geared towards learning beween different actors, including, for example, the state, donors, non-governmental organizations, local groups, researchers and activists. If such urban forums are often sites of exclusion, managerialism and control, they also embody the historical potential of learning between constituencies to develop not just more democratized urbanisms, but more socially just urbanisms.

Dwelling and Perception

Translation and coordination are concepts that provide an insight into how learning is produced and how it operates, but we have said relatively little so far about how learning is *lived*. It is in this context that I use the notion of dwelling to consider how learning emerges though relations between individual or group and the city. If dwelling has experienced some theoretical resurgence in geography (e.g. Elden 2001; Obrador-Pons 2003; Harrison 2007; Jacobs and Smith 2008; Kraftl and Adey 2008; McFarlane 2011a), there has been little attempt to connect dwelling to learning specifically. The work of anthropologist Tim Ingold, particularly in his (2000) book *The Perception of Environment*, is an importance exception here. Ingold (2000) examined learning in relation to skill and dwelling in the premise that people are always part of the process of coming-into-being of the world. From this perspective, a process like urban policy production occurs through attuning perception to sites, documents and events in a process of immersion. This immersion, which Ingold (2000: 154) called a 'dwelling perspective' inspired by Heidegger and phenomenology, insists that worlds are made, whether in imagination or 'on the ground', 'within the current of their life activities'. One implication is that meaning, for instance in relation to an urban policy, is 'immanent in the context of people's pragmatic engagements' with the document, environs, discourse or idea; meaning is located in the relational contexts of people's 'practical engagement with their lived-in environments' (Ingold 2000: 154, 168). As Obrador-Pons (2003: 49) wrote of the Heideggerian *dasein* ['being-there'], *dasein:* 'is always already amidst-the-world. Our involvement, that is, our way of dwelling in the world is mainly practical not cognitive. Being-in-the-world is an everyday skilful, embodied coping or engagement with the environment.' This means that people learn to perceive policy through a practised ability to notice and to respond to changing contexts: the ways in which we know, learn, coordinate, build and negotiate depend not just on the translation or coordination of knowledge, but on what Ingold called, after psychologist James Gibson, an 'education of attention' (Gibson 1979: 254; Ingold 2000: 166–7; Seamon 1993, 1998, 2000).

In this education of attention, learning through dwelling entails shifts in perception, a way of seeing that is haptic – sensed, embodied, practised – and which positions learning as a changing process of perceiving how to use the affordances of documents, objects and situations. What matters most about dwelling, as Heidegger (1971) suggested in relation to housing, is that people must *learn* to dwell. Perception creates knowledge that is practical because it is based on whatever activity the person is currently engaged in: 'to perceive an object or event is to perceive what it *affords*' (Ingold 2000: 166; emphasis in original). As Lingis (1996: 14, cited in Harrison, 2007: 631) argued through Heidegger: 'To see something is to see what it is for; we see not shapes but possibilities.' Or, as Hinchliffe *et al.* (2005: 648) wrote in relation to how their perception of the landscape shifted in their research in urban wildlife: 'As Latour (2004) might say, we had started to learn to be affected. We were bodies in process, gaining new ways of looking, a new set of eyes (or newly conditioned retina), a slightly more wary nose, a different sensibility.' Learning, then, involves not just technical competence, but developing forms of relatedness to objects. The world, argued Ingold (2008: 1797) in a later essay, is not just occupied by already existing things, but *inhabited*, i.e. 'woven from the strands of their continual coming-into-being'. We might consider, for instance, how urban infrastructure comes to matter through particular uses and practices, as Susan Leigh Star (1999: 380) argued in relation to water infrastructures: 'The cook considers the water system as working infrastructure integral to making dinner. For the city planner or the plumber, it is a variable in a complex planning process or a target for repair.'

We are not particularly far away here from Bourdieu's (1977) work on *habitus*: the ways in which people learn specific dispositions and sensibilities over time that are particular relations to their environment, thereby enabling and inhibiting different kinds of learning and action. Ingold (2000) traced this education in the production of skill as a practice-based form of fine-tuning. There is a set of ethico-political concerns here around how people learn to dwell in the city, for example, in whether people choose to participate in fair trade and recycling, whether they practise a hospitable openness to difference as cosmopolitan dwelling, or whether and how they are involved in local activism or political engagement.

There is a range of different modes of learning through dwelling that we might identify. Amin and Roberts (2008a: 366), writing about organizational learning, identified several: craft/task knowing, involving repeated practice and close supervision; professional knowing, involving internalizing codified knowledge while developing a 'feel' for the professional habitus; epistemic/creative knowing, based less around a sense of community than a shared problem; and virtual knowing, often driven by enthusiasts with weak or anonymous personal ties. Importantly, they argue that the success of organizational learning cannot be reduced to spatial proximity: '... what determines the texture of ties or trust is not spatial proximity, but the nature of contact, intermediation, and communicative complexity involving groups of actors and entities' (Amin and Roberts 2008a: 366; 2008b). This emphasis on process rather than proximity disrupts the often bounded, 'earthy' baggage of connotations that dwelling carries – the danger in the concept of dwelling that Ingold himself has acknowledged of a stubborn connotation of 'snug, well-wrapped localism' (Ingold, 2008: 1808; and see Hinchliffe 2003; Rajchman 1998). But in unsettling the spatiality of learning through dwelling, we need to be equally mindful of the multiple temporalities of dwelling. While dwelling focuses attention on how learning emerges through engagement with the everyday city, this does not mean that learning is reducible to the present moment. In dwelling the city, people draw upon previous experience or memories, and the multiple temporalities and rhythms of the city itself help to shape the possibilities of learning through dwelling, from rhythms of day and night, to capitalist cycles of growth, collapse, shrinkage and decay, to the rhythms of long-term migration.

We are left, then, with a theory of learning based on three interrelated ongoing processes: translation, or the relational distributions through which learning is produced as a sociomaterial epistemology of displacement and change; coordination, or the construction of functional systems that enable learning as a means of coping with complexity, facilitating adaptation, and organizing different domains of knowledge; and dwelling, or the education of attention through which learning operates as a way of seeing and inhabiting urban worlds. These three areas are inextricably co-constitutive and dependent; if learning involves the translation of knowledge across space-times, it must none the less be coordinated, and if learning requires organizational devices, it must none the less be lived through everyday practices of dwelling. But, despite the constitutive presence of spatiality to each of these three concepts, this schema of learning as translation–coordination–dwelling falls short of providing a theory of the spatialities of learning. In the next section, I foreground the spatialities of learning by asking what sort of spatial grammar might be used to conceptualize the spatialities of urban learning. It is in this context that I use the concept of assemblage.

Assemblage Space

We can think about assemblage in a variety of ways. There is a general usage of assemblage as a description of how different elements come together. This sense of assemblage contrasts with a more explicit rendering of assemblage as a concept for describing unity across differences – a name for relations between objects that make up the world through their interactions. And there is assemblage as an approach, an orientation that operates as a way of thinking of the social, political, economic or cultural as a relational process of composition, and as a methodology attuned to practice, materiality and emergence. These are not necessarily mutually exclusive positions; we can think of assemblage as both orientation to the world (e.g. a form of thinking about urban policy production) and as an object in the world (e.g. an urban policy, house, or infrastructure). In offering a conception of learning as assemblage, I am working with both these senses of assemblage as orientation and assemblage as object. But, as part of both these positions, I am also thinking of assemblage as broadly political – as a way of thinking about not just how learning is produced, but how cities might be learnt differently, i.e. assemblage as a means of continually thinking the play between the actual and the possible (Chapter 6; McFarlane 2011b).

As a general currency, assemblage is increasingly used to connote, expansively, indeterminacy, emergence, becoming, processuality, turbulence, and the sociomateriality of phenomena. It is, then, part of a more general reconstitution of the social field as materially heterogeneous and practice-based (Schatzki 2001; Latour 2005; Massey 2005; De Landa 2006; Thrift 2007; Anderson and Harrison 2010). As a descriptive term for transgressing modernist dualisms like nature-culture, body-technology, or physical-political, it often functions as a *style* of knowledge production alert to compositional alignment and realignment (Phillips 2006). In urban geography, assemblage has been deployed in various ways: as a descriptor of sociomaterial transformation in accounts of urban socionatures, cyborg urbanisms, or urban metabolisms (e.g. Gandy 2005; Swyngedouw 2006); as a means for thinking through the contribution of actor–network theory for rethinking the city, for example in Farías's (2009) usage of assemblage as a basis for decentring the city and rendering urbanism as a multiplicity of processes of becoming, sociotechnical networking, and heterogeneous collectivity, building on Deleuze and Guattari's (1981) notion of *agencement* – the alignment of different elements; and in relation to urban policy mobilities, including the relations between travelling policies and their localized substantiations (e.g. Allen and Cochrane 2007, 2010; Ong 2007; Sassen 2007; Farías and Bender 2009; McGuirk and Dowling 2009; McCann and Ward 2011).

While I would resist the temptation to view the disparate usages of assemblage as a historical common field, there are, as Venn (2006a: 107) argued, a set of emphases that many uses of the concept share, including 'adaptivity rather than fixity or essence ... co-articulation and compossibility rather than linear and discrete determination ... and the temporality of processes'. Assemblage, then, has become a vocabulary for describing the productivist alignment of different sources, but is rarely itself an object of conceptual elaboration. As a general working definition I use the Deleuzian conception of assemblage as 'a multiplicity constituted by heterogeneous terms and which establishes liaisons, relations between them' (Deleuze and Parnet 2007: 52). For Deleuze, the only unity of assemblage is that of 'co-functioning: it is a symbiosis, a "sympathy". It is never filiations which are important but alliances, alloys; these are not successions, lines of descent, but contagions, epidemics, the wind' (ibid.). This means that urban actors, forms or processes are defined less by a pre-given definition and more by the assemblages they enter and reconstitute. To take an example that will feature in Chapter 5, nineteenth-century urban sanitation infrastructure made possible new discourses of public health and practices of private hygiene – the component parts were constituted through the interaction between infrastructure, discourse and practice. Or, to take a different example, urban policy recommendations might depend upon learning particular statistical knowledges about specific domains of urban life, such as health or education – again, these assemblages exist through their interaction.

The individual elements define the assemblage by their co-functioning, and can be stabilized (territorialized or reterritorialized) or destabilized (deterritorialized) through this mutual imbrication. But this is not to say that an assemblage is the result of the properties of its component parts. It is the *interactions* between components that form the assemblage, and these interactions cannot be reduced to individual properties alone. As a form of spatial relationality, assemblage thinking is attentive to both the individual elements *and* the agency of the interactive whole, where the agency of both can change over time and through interactions. Rather than being exhausted by specific interactions, the parts may provide future resources for, or be altered by, the assemblage. The changing nature of assemblages through interactions is one of the ways in which, as Manuel De Landa (2006: 10–11) has argued, assemblages operate as wholes characterized by 'relations of exteriority'. The other sense in which assemblages are characterized by relations of exteriority is that component parts may be detached and plugged into a different assemblage in which its interactions are different. Learning is an assemblage constituted through interactions that emerge through processes of translation, coordination and dwelling. As a spatiality of learning, assemblage departs from three broad starting points.

Firstly, as an orientation, assemblage is an attempt to emphasize that urban learning is constituted by relations between history and potential, or the actual and the possible. Here, I am referring both to the crucial role of urban histories in shaping trajectories of urban policy and economy, habits of practice, and ways of going on – and therefore the context for urban learning – and to the potential and excessiveness of the moment, the capacity of events to disrupt patterns, generate new encounters with people and objects, and invent new connections and ways of inhabiting everyday urban life and therefore to create different possibilities for learning. Potential signals the relation between the actual and the possible in learning – between the city that is known and the city that might be or could have been – and thereby speaks both to the urban imagination, the sense of possibility that the city can generate under varying conditions of restraint and inequality. As Nicholas Tampio (2009: 385) has argued, assemblage, for Deleuze, was oriented towards *actualizing* ideals of freedom and equality. As a means for thinking the spatialities of urban learning, assemblage draws attention not just to an ecology of relations, but more to the *particular urban alignments* formed through the multiple spatiotemporalities of translation, coordination and dwelling. At particular moments and for certain durations, different exteriorities can enter into the constitution of assemblages, only to change or disperse at a different time or from a particular 'angle of vision' (De Landa 2006; Li 2007). If elements of this reading of assemblage connect with certain conceptions of network, it is worth highlighting the broad contrasts between assemblage and the notion of network.

In relation to actor–network theory (ANT), for example, assemblages are relations not just of stability and rigidity, but of excess, flux, and transformation. This is not to say, of course, that ANT studies only emphasize rigidity and stability, but that the emphasis is often on these forms. As Ong (2007: 5) argued in relation to assemblage and neoliberalism:

> Although assemblage invokes nexus, it is radically different from concepts such as 'network society' or 'actor network theory' that seek to describe a fully fledged system geared toward a single goal of maximization. ... The space of assemblage is the space of neoliberal intervention as well as its resolution of problems of governing and living.

Assemblage connotes transformation, or the work of reassembling, thereby focusing attention on the possibility of invention and potential. Invention here operates not necessarily as something new but, as Barry (2001: 211–12) has argued, arrangements in which objects or devices become:

> ... aligned with inventive ways of thinking and doing and configuring and reconfiguring relations with other actors. ... What is inventive is not the novelty of artefacts and devices in themselves, but the novelty of the arrangements with other objects and activities within which artefacts and instruments are situated, and might be situated in the future.

While for ANT the network delimits focus to the interaction between individual parts, assemblage is more attentive to the changing agency of both the parts and the alignments. Assemblage implies a greater conceptual openness to the unexpected outcomes of disparate intentions and activities.

However, the conception of assemblage offered here is not one that is intended to function as an alternative to the notion of network in actor–network theory. Indeed, if ANT has often been preoccupied with the stabilization of networks, the focus has been increasingly on an 'ethos stressing fluidity, transformation and ambivalence' (Van Loon 2006: 310). But as Legg (2009) points out, even ANT's main protagonist, Bruno Latour, has distanced himself from the persistent tendency to connote network with rigidity and to undermine the complexity of relations such as structure/complexity or human/non-human. If assemblage differs from some ANT readings of network in that it attends both to change and rigidity, it none the less exists in similar conceptual terrain, attempting to confront the complexity of sociomaterial relationality.

Secondly, as a conception of urbanism, assemblage emphasizes how learning is produced not as simply a spatial category, output, or resultant formation, but as a process of doing, performance and events. There is no necessary spatial template for assemblage; the spatiality of assemblage is that of sociomaterial alignment, which brings into view a range of spatial forms, from those generated by historical processes of capital accumulation and social polarization, to random juxtapositions and disruptive events, and predictable daily and nightly rhythms of activity, atmosphere and sociability. Although the specific character of assemblage is a product of interactions, assemblages are not reducible to events or practice, but must be understood in the context of their historical production and transformation. At different moments of time, learning within and between sites and actors may require different kinds of labour and are more or less vulnerable to collapse, or to reassembling in different forms. As Bennett (2005: 461) pointed out, drawing on Deleuze and Guattari (1987), this underlines the agency not just of each member of the assemblage, but of the groupings themselves: the milieu, or specific arrangement of things, through which forces and trajectories inhere and transform. The different 'parts' of a learning assemblage do not interact atomistically but as co-constituting relations that define one another. For all that this underlines the spatiotemporality of assemblage, it is worth noting here that for some Foucauldian scholars the temporality of assemblage is that of the ephemeral rather than the *longue durée* (e.g. Rabinow 2003; Li 2007; Legg 2009). For Rabinow (2003: 56), assemblage is an 'experimental matrix of heterogeneous elements, techniques and concepts' that disappears in years or decades rather than centuries. Longer lasting 'problematizations' – what he calls 'grander problematizations' – connote less a sense of emergence and more a sense of resultant formation, and their form is that of the apparatus or *dispositif* (see Li 2007; Legg 2009). In contrast, while assemblage does emphasize emergence, I do not assign any particular temporality to assemblage, nor any necessary level of stability. Rather than opposing assemblage to apparatus, I prefer to think of how forms of power, rationality and intelligibility structure and enclose urban assemblages, or – to use a Deleuzian language of assemblage – territorialize, deterritorialize or reterritorialize (De Landa 2006; Dovey 2010). Assemblages can be made singular through the action of particularly powerful agendas or groups, even if there is always, to varying extents, the potential to be otherwise. Elements are drawn together at particular conjunctures only to disperse or realign. What this begins to outline is a conception of urban learning assemblages of actual and potential urbanisms located in emergent material practice, shaped by trajectories of urban history, and which are not characterized by any necessary pre-given spatial or temporal templates.

Thirdly, and finally, urban learning assemblages are sociospatially structured, hierarchalized and narrativized through profoundly unequal relations of power, resource and knowledge. Rather than a kind of crude opposition to structural hierarchy, the spatialities and temporalities of urban learning assemblages – for instance in relation to policy or development formations – can be captured, structured and storied more effectively and with greater influence by particular actors or processes than by others. As the examples cited above of Gandy (2005) and Swyngedouw (2006) have in their different ways vividly demonstrated, power, political economy and sociocultural exclusion are central to how urban assemblages are produced. For example, Gandy (2005) showed how the cyborg figure allows a critical purchase on connections between body, technology, exclusion and violence. Whether in the functioning – or non-functioning – of infrastructures as life-support systems; or in the sociomaterial militarization of society through the technologically-enhanced urban soldier, the destruction of civic infrastructure, or the radical extension of surveillance technologies through cities (Graham 2008, 2009); or in the 'decyborginization' to bare life of the marginalized through violence, impoverishment and disease, Gandy (2005: 32) showed how certain forms of urbanism have the power to destroy, reduce and enable particular forms of urban life. Swyngedouw (2006), in his critical elucidation of urbanization as the de-territorialization and reterritorialization of metabolic flows, argued that unequal relations of power allow particular actors to defend and create their own urban environments along lines of class, ethnicity, race and gender. As he wrote (2006: 106):

> Under capitalism, the commodity relation and the flow of money veils and hides the multiple socioecological processes of domination/subordination and exploitation/repression that feed the urbanization process and turn the city into a metabolic socio-environmental process that stretches from the immediate environment to the remotest corners of the globe.

As I hope to show throughout the book, and examine in detail in Chapter 6, assemblage can serve as a conceptual tool for illuminating a critical geography of urban learning.

Assemblage focuses on emergence, and in doing so drives critical attention to why and how particular forms of urban learning become dominant over others, for example, why and how certain forms of knowledge travel while others are marginalized. For instance, in relation to urban policy, recent years have witnessed a dominance of very particular forms of 'creative' or 'smart city' – particular images of the learning city – which are made to travel at the expense of others (Florida 2002, 2005; Peck 2005). Hollands (2008) attempted to recuperate the discourse of 'smart cities' from what he saw as its elitist and exclusive imaginaries, and considered what an alternative conception of the travelling 'smart city' might look like. In outlining a progressive and inclusive smart city, Hollands (2008: 312) asked how do discourses of the 'smart city' 'relate to the "less" smart/creative sections of the population? ... [W]hile smart cities may fly the banner of creativity, diversity, tolerance and culture, the balance appears to be tipped towards appealing to knowledge and creative workers, rather than using IT and arts to promote social inclusion.' In response, he provisionally offered two urban learning assemblages that a progressive smart city might strive for. First, he argued, any notion of the progressive smart city must start with people and their existing knowledge and skills, rather than beginning with technology. This requires positioning information technology to 'empower and educate people, and get them involved in a political debate about their own lives and the urban environment that they inhabit' (Hollands 2008: 315), an imperative that entails working hard to involve different people in both the role and the activities of information technologies. Secondly, there needs to be a shift in the balance of power between the use of technology by business, government and communities, to provide more opportunities for enhancing citizen participation and in local decision-making. In this sense, information technology (IT) could potentially be used to facilitate a 'virtual public culture' (ibid.). Rather than allowing the smart city label to mask underlying urban inequalities, then, Hollands (2008: 316) seeks to expose and evaluate smart city discourse and to offer an alternative conception of smart cities that 'take much greater risks with technology, devolve power, tackle inequalities and redefine what they mean by smart itself'. ...

In his account, Hollands discussed the actual and potential urbanisms of 'smart cities' and set them against an alternative urbanism that is inclusive and people-centred. The specificity of assemblage here lies in this relation between the actual and possible city: as a mode of critique, assemblage is both an *analytic* that continually asks 'how have particular forms of urban learning become dominant and made to travel, and how might urbanism have been produced differently?', and an *orientation* towards alternative forms of urban learning. In focusing on urban production, assemblage offers one route for thinking through urban potentialities – alternative, more socially just urbanisms that might otherwise emerge and travel. Many of the urban learning assemblages I encounter in this book are translocal in nature, and while all assemblages are to varying degrees translocal, I write of 'translocal assemblage' to underline this relational spatiality. I purposively use 'translocal' rather than 'global' here.[1] This contrasts, most obviously, with Ong and Collier's (2005) influential edited collection, *Global Assemblages*, which focused on the articulation of 'global forms' as territorialized assemblages (where 'form' refers to abstractable technologies as different as neoliberalism, citizenship, democracy, or international regulations). For Ong and Collier (2005: 4), assemblages are material, collective and discursive relationships, and in focusing on the specificities of global forms in particular sites they are interested in the formation and reformation of assemblages as political and ethical 'anthropological problems'.

---

[1] There are broad echoes here of M.P. Smith's (2001: 17, 19) reading of 'translocal' as 'multi-sited' formations of 'social actors engaged in a reterritorialized politics of place-making', although my focus, as I will make clear, is specifically on the idea of urban learning assemblages (see also Clifford 1997: 7).

In an important passage, Ong and Collier (2005: 12) clarify the relation between global form and assemblage, including the question of their spatial templates:

> In relationship to 'the global', the assemblage is not a 'locality' to which broader forces are counterposed. Nor is it the structural effect of such forces. An assemblage is the product of multiple determinations that are not reducible to a single logic. The temporality of an assemblage is emergent. It does not always involve new forms, but forms that are shifting, in formation, or at stake. As a composite concept, the term 'global assemblage' suggests inherent tensions: global implies broadly encompassing, seamless and mobile; assemblage implies heterogeneous, contingent, unstable, partial and situated.

This passage is a useful specification, particularly in its emphasis on assemblage as a composite, contingent and emergent concept. And yet, despite their stated intention of avoiding characterizing forms as 'global' and assemblages as 'local', assemblage is substantiated in this account as a set of 'reflective practices' through which global forms are subjected to critical questioning. In this move, the distinction between 'global' and 'local' resurfaces. It is in this context that I am using the prefix 'translocal' as an attempt to emphasize the blurring of that scalar distinction in the production of urban learning assemblages (and in this sense, as a distinction from the scalar distinction of assemblage found in De Landa's (2006) ontology, *A New Philosophy of Society: Assemblage Theory and Social Complexity*.

As a compositional unity – not necessarily of internal coherence but of elements aligned together (Li 2007; McGuirk and Dowling 2009) – assemblages are constituted by relations of history and potential, and of doing and performance, and are structured through unequal relations of knowledge, power and resource. As a spatiality of learning, assemblage focuses attention on how translation, coordination and dwelling are spatially composed and aligned. Assemblage is attuned to the emphasis with translation on sociomaterial spatial distributions, on the territoriality or enabling capacities of coordination, and on dwelling's focus on the immersion and affordances of everyday urbanism.

Conclusion

The critical purchase of the concept of urban learning assemblage is not simply a call to know more of cities, but to unpack and debate the politics of learning cities by placing learning explicitly at the heart of the urban agenda. This learning and relearning is based on three processes: translation, or the relational and comparative distributions through which learning is produced as a sociomaterial epistemology of displacement and change; coordination, or the construction of functional systems that enable learning as a means of coping with complexity, facilitating adaptation and organizing different domains of knowledge; and dwelling, or the education of attention through which learning operates as a way of seeing and inhabiting the world. I have outlined a notion of assemblage as a spatial grammar of urban learning focusing on the constitution of learning through relations of history and potential, doing and performance, and structured (although not delimited by) inequalities. Assemblage connotes the processual, excessive, and performative; it is a conception of the spatialities of urban learning as an unfolding of distinct sociomaterial rationalities and processes through emerging and unequal milieu. Urban assemblages are not simply a spatial category of learning, nor are they outputs or resultant formations. Instead, assemblage focuses on how learning operates through doing, performance and events. At different moments in time, particular forms of urban learning may require different kinds of labour and are more or less vulnerable to collapse, or to reassembling in different forms.

While this chapter has developed a conceptualization of learning assemblages, it is only through subsequent chapters that particular manifestations of urban learning assemblages will emerge. As Olds and Thrift (2005: 202) have written in a different context, assemblages will function quite differently across different contexts, 'not because they are an overarching structure adapting its rules for the particular situation, but because these manifestations are what the assemblage consists of'. The next chapter begins this process of applying urban learning assemblages to urbanism by charting a range of ways in which everyday urban life is produced by, demands, and is contested through different forms of urban learning.

Chapter Two

Assembling the Everyday: Incremental Urbanism and Tactical Learning

Introduction

For most urbanites, urban learning occurs through social and spatial practice in the city. Learning the city emerges not through a formal, linear cognitive process, but through experiential immersion in urban space-time. And yet, there has been surprisingly little attention to urban learning amongst ordinary people and groups in the city (for exceptions, see Chattopadhyay 2009; Hansen and Verkaaik 2009; Roy 2009a,b; Simone 2004a; Shove *et al.* 2007). This chapter focuses more on forms of urban dwelling than on translation and coordination, although the three remain inextricably interrelated. In doing so, it offers two interrelated conceptualizations of how learning features in the production and contestation of everyday urban life: *incremental urbanism* and *tactical learning*. These processes are particular modes of learning through dwelling, and manifest in a range of ways, whether as a form of accretion – i.e. a cumulative and shifting set of practised knowledge, rather than a kind of linear addition – or as more defined and sudden shifts in perception or ways of experiencing urbanism.

To take just three important and common sets of urban experience that will run through the chapter: various *forms of improvisation*, sometimes voluntary and sometimes forced, including domains as different as housing or infrastructure, to systems that deal with water shortages, to pastimes like urban skateboarding, inform changing conceptions and experiences of the city; *everyday rhythms*, from different forms of walking and mobility, to work routines, predictable patterns of activity, and the negotiation of different rhythms, are central to how the city is inhabited; and a range of *coping*

Learning the City: Knowledge and Translocal Assemblage, First Edition. Colin McFarlane.
© 2011 Colin McFarlane. Published 2011 by Blackwell Publishing Ltd.

mechanisms and strategies for making a living or negotiating the city are important to how people respond to urban processes and forms through learning. These are largely, although not exclusively, everyday experiential forms of learning the city which any urban dweller can relate to, but which are experienced in radically different ways through relations of class, gender, race, age, religion, and so on. As Borden *et al.* (2001: 9) reflected in their introduction to *The Unknown City*, learning the city is first and foremost a 'project of becoming' though unfolding events and struggles over time and space, the city as 'a conjunction of seemingly endless possibilities of remaking' (Simone, 2004b: 9).

These different forms of improvisation, urban rhythms, and coping mechanisms are examples of learning through everyday urbanism that both emerge through and are productive of incremental urbanism as a crucial process of dwelling, and which can operate as urban tactics, i.e. intentional, proactive attempts to advance in the city through learning. A focus on the everyday, as Edgar Pieterse (2008: 113) has argued, opens the possibility of locating learning within a larger matrix of urbanism often missing from accounts of the city – the 'below-the-radar' sets of small actions to, for example, gain access to infrastructure or to extend housing space. I argue that different forms of learning are crucial to how different groups address urban marginality, and that these can be usefully conceptualized as ongoing forms of urban tactical learning. In doing so, I consider how forms of tactical learning can interrupt and widen the possibilities of urban dwelling. Throughout the chapter I will draw predominantly on illustrative examples from fieldwork conducted in Mumbai and, in the section that follows, São Paulo.

Incremental Urbanism

In an afternoon in August 2008, I accompanied Gabriella, a social worker, as she visited project sites in São Paulo's large favela, Paraisópolis. Paraisópolis is a settlement of some 70,000 people located on the steep hills of southern São Paulo. Adjacent to the settlement is Morumbi, an élite gated suburb that towers above the red-brick neighbourhood (Caldeira, 2000). Late in the afternoon, Gabriella took a detour to show me what she described as an entirely unique house. She was not exaggerating. The man who came to the spectacular arched doorway knew Gabriella, and narrated the story of how he built the house through amalgamating coloured stone and just about any object that came to hand, constructing the house from the sea of everyday life. The roof was made up of, amongst other things, discarded pieces of plastic, old shoes, an iron, children's toys, mugs, kitchen utensils, and all manner of other materials, cemented into an intricate arrangement of decorative, coloured stones. He patted the walls and ceiling

Figure 1 House, Paraisópolis, São Paulo, used by permission of the author

proudly and gestured over his shoulder, revealing an interior that was even more impressive than the exterior – a rabbit warren of smoothed and angular stones and archways assembled through a wide range of materials, generating a prism of light and dark as the sun shone here and there. On the roof is a garden used by his children that continues the theme. The house was in constant construction, he said, and the project had taken years so far. Figures 1 and 2 show some of this complex assortment.

I should say at this point that I do not wish to romanticize this house. The house, creative as it is, is nonetheless situated within a poor, precarious neighbourhood where people struggle to make a safe and sustainable livelihood and are sometimes victims of appalling abuse and violence. I want to use this example as a heuristic that reveals how learning as dwelling occurs through a process of incremental assembly. The house focuses our attention

Figure 2 Ceiling of house, Paraisópolis, São Paulo, used by permission of the author

on urban materialism, a materialism of the *things themselves*: what materials might be used for, how they might function out of place, whether they are 'useful' in this context, and the sheer improbability of their juxtaposition (see Bennett 2010). Our attention focused, these materials are none the less made strange. In this urban uncanny, the materials relate only because of their alignment as part of the roof or walls of the house. In seeing these materials used unexpectedly, we are both reminded of their more typical functions (e.g. that of an eggbox or an iron) while clear that they are defined by their new context as interrelations that constitute part of the roof. In short, the individual artefacts are defined by a potentiality that exceeds their more familiar and usual functions, while also constituting a new assemblage.

The materiality of the house also casts up another immediate sense, then: that of the builder and his or her own motivations and skills. The house was built by an artist who has designed other areas of Paraisópolis, including a public stairwell for the settlement in a style reminiscent of Gaudi's colourful non-linear designs in Barcelona. We are presented, then, with a sense of the labour and craft that has been put into this house. The seemingly random assortment of materials suggests that the construction was built through embodied immersion rather than abstract contemplation. In other words, the very materiality at work in this house signals a form of learning that is a haptic rather than purely cognitive or optical process. This is a form of learning through assemblage in that the materials are defined by their new relation with one another, i.e. learning takes place by seeing not just materials, but *possibilities* – in this case of bringing materials into new interactions. What this opens up, I suggest, is the question of how human–material relations function in the urban environment, and what these relations reveal is the urban as a process of learning through incremental urbanism – of the city as multiple, unfolding habitation.

Housing is often an incremental process of gradual manipulations within the urban environment – 'tinkering', to borrow from Anne-Marie Mol (2008), housing-as-a-verb, in John Turner's (1972) memorable description. As literature on the material culture of the home has demonstrated, home-making in its most general sense is a cumulative process of assembly (e.g. Noble 2004; Jacobs and Smith 2008). As Jacobs and Smith (2008: 517) argued, the house is a set of contingent sociomaterial orderings, the constitutive geographies of which extend far beyond the territory of the house:

> The acts of 'housing' and 'dwelling' are a coproduction between those who are housed and the variant technologies that do the work of housing: ornaments and decorations, yes, architecture and bricks and mortar, sanitation and communication technologies, too, but also housing policies and practices, mortgage lending and insurance, credit scores, and all the other lively 'things' of finance.

For Jacobs and Smith (2008: 518), part of the challenge here is to do away with the home/housing binary in order to focus more upon the dispersed logics, practices, meanings and experiences that perform 'home' as an 'assemblage of dwelling'. In this process of assembly, different 'lively things' are learnt through their interrelation with one another. In one of Heidegger's (1971) descriptions, dwelling is not just about engineering, architecture or *techne* (in the Greek conception of 'letting appear'), but the raising of locations and joining of their spaces through gathering, or assembly. But as the example from Paraisópolis suggests, the learning of dwelling is structured by stark geographies of inequality.

For example, one non-governmental agency working on low-income housing based in Mumbai, where I do most of my research, describes a process of incremental housing that takes place in the context of often severe urban poverty (Indian Alliance 2008: 6). In this passage, they describe a family of migrants who have just arrived in the city and who have found a location for housing:

> They start out by putting up plastic sheets on poles, under which they sleep at night and pack up during the day. Over time, corrugated metal sheets replace the plastic, which become the walls and ceiling of the shack, to be later

replaced by bricks and mortar. ... Gradually, the roof becomes the first floor, as metal sheets are put up as walls. These eventually become concrete as well, with ladders or narrow, steep staircases leading up from the outside. ... Additional floors are often rented out to other migrants, thus increasing the income of the original family ... slum dwellers continue building these incremental levels. (Indian Alliance 2008)

This trajectory of incremental housing does not, of course, apply to all people living in poor neighbourhoods in Mumbai, let alone anywhere else, and the trajectory itself is mediated by class, caste, religion, ethnicity, gender, and other factors. The narrative also hides the various actors that the poor must negotiate in order to get access to different infrastructures and services, not to mention the labour and costs involved. It masks the multiple processes through which, as AbdouMaliq Simone (2008a: 13) has written of African cities, 'actors come to make their mark on collective transactions and the way in which idiosyncratic constellations of such actors provide a workable balance between the provisional and incessantly mutating practices required to viably "make do" in most African cities and a sense of order, if only temporary'. In other words, this narrative underplays the ways in which people *inhabit the learning assemblage*, i.e. how they live through and learn to negotiate the varying forms of porosity and closure of the assemblage, including the possibilities that assemblage opens but which are not part of its current alignment. This notwithstanding, the central point that I want to make here is that of learning as a process of incremental *dwelling as assembly*.

As Chapter 1 argued, learning through dwelling is also dependent upon forms of translation and coordination. The process of sociomaterial engineering described above involves the translation of various materials into new uses over time, including roofs that become, first, floors for sleeping, then later, spaces for renting out; or ladders that shift from being access points to the roof to stairs for a new family living in a newly-built shack on the roof. In each step in this 'chain of translation' (Latour 1999), materials – such as the ladder, or the corrugated metal sheets – operate as functional systems of learning through the coordination of different domains, from the limited spaces of the shack to the aspirations and desires of the inhabitants and the availability of money and materials – an everyday architecture embedded in the routines, activities and temporalities of quotidian life, 'part of the flow of space and time, part of the interproduction of space, time and social being' (Borden *et al.* 2001: 11). What is being learnt incrementally here is the different affordances of materials over time, i.e. learning as a different way of seeing urban materials, as an education of attention to the urban environment.

In this context, dwelling underlines how the perception of objects pertains to what objects might offer in the context of a precarious and impoverished everyday life – as Cullen and Knox (1982: 285) argued in a searching paper on the self and the city: '[W]e do not discover the "usability" of things by observing them, or by observing them and establishing their properties – but by the 'circumspection of the dealings in which we use them' (Heidegger, 1962: 102). We learn urban space by seeing not just materials but the possibilities of those materials through particular contexts. These 'learnt objects' are not just practically useful because of what they might afford for new relations, they also play an active part in conditioning the possibilities of urban life. Dovey (2010) describes this processual multiplicity through a distinction between extensive and intensive multiplicity, drawing on Deleuze and Guattari (1987). An 'extensive multiplicity' is where the constituent parts are defined by their spatial extension and are unaffected by new additions, whereas an 'intensive multiplicity' is a multiplicity that can, to different extents, be changed by new additions: 'A house, neighbourhood or city is an intensive multiplicity. When different people move in, new buildings or rooms are added, the sense of the larger place changes' (Dovey 2010: 27). While the shack is *built*, then, it is a practice more accurately described as *dwelt*. As habitat, the incremental shack is made through the current of precarious life activities. Notwithstanding its pejorative connotations, it is in this sense that the ubiquitous term 'slum dweller' resonates. But this dwelling also offers a window into the city more generally: the city as an ongoing chain of sociomaterial translations that affords different possibilities at different times under deeply unequal conditions for urban learning.

Incrementalism, as laborious and historical accretion, is central to learning through dwelling, and is common to a whole range of urban processes and forms, from housing and policy to infrastructure and culture. As Simone (2008a: 28) has argued, cities:

> ... no matter how depleted and fragmented, still constitute platforms for trajectories of incrementalism. Houses and limited infrastructure are added onto bit by bit; the mobilization of family labor buys time for a small business to grow; migration is used as an instrument to pool together savings in order to start a new economic activity; mobile work crews are formed to dig wells, help with construction, or deliver goods until they make enough contacts to specialize on one particular activity.

Incrementalism is not a process of linear addition. Learning is a process of translation, meaning that knowledge about one issue can be used in different contexts, or perhaps might be forgotten altogether. Moreover, incremental learning is not the production of an unmediated stockpile of knowledge; these forms of urban learning are stratified, unequal and controlled. For instance, some groups, Simone (2008a: 28) continues, 'are able to organize labor, money, and contacts to finish roads, complete water reticulation projects, or electrify their compounds and neighborhoods, while others languish'. The livelihoods of millions of people living in marginalized settlements are constituted by a range of assemblages and strategies produced on an everyday basis, and 'through such delicate negotiations the urban fabric of the city, a constellation of irregular forms, becomes legitimised' (AlSayyad and Roy 2006: 10). We see this continuous effort, for instance, in the makeshift urbanism that constitutes much of the daily life for people within marginalized informal settlements, especially women who tend to take on the majority of household construction and maintenance, as Neuwirth has written: 'With makeshift materials, they are building a future in a society that has always viewed them as people without a future. In this very concrete way, they are asserting their own being' (Neuwirth 2006: 21–2). This makeshift urbanism, what De Certeau (1984: xv) may have called bricolage, refers to 'ways of "making do"', and connotes a key form of incremental urbanism that is often used in descriptions of urban informality: *improvisation*.

The production of incremental housing through learning different uses and relations of materials over time demonstrates that, contrary to how informal settlements are often perceived, improvisation is not simply *ad hoc* but the product of tinkering and tweaking through urban assemblages (see Dovey 2010). This is the case for cities more generally, as Graham and Thrift (2007) have shown in relation to the forms of hidden learning and adaptation that constantly occur on a daily basis in the maintenance and repair work of urban infrastructures, from roads and electricity to water and sanitation. As they argue, it is through these often labour-intensive sociotechnical processes of maintenance and repair that the urban world often appears to us 'ready-to-hand'. Here, I see improvisation not as sudden shifts, but as a process of creatively tinkering with urban space that emerges through incremental learning. To take a quite different example, Borden (2001: 196) shows how over time, urban skateboarders learn to improvise with urban objects in different ways through practical immersion – walls, benches, pavements, steps and gaps can take on a 'newness' as they are creatively recast as objects of play or exploration. Borden shows how skateboarders learn to use everyday urban materials in different ways by seeing new possibilities in those materials. For instance, a public handrail is transformed from being an object of safety to becoming an object of risk as skateboarders jump on to and slide down it. Improvised learning here involves a simultaneous and co-constituting learning of both urban microspaces and the craft of skateboarding gradually over time, as Borden (2001: 190) relates in relation to sound and touch:

> [T]he sound of the skateboard over the ground yields much information about the condition of the surface, such as its speed, grip, and predictability ... a sense of touch, generated either from direct contact with the terrain – hand

on building, foot on wall – or from the smoothness and textual rhythms of the surface underneath, passed up through the wheels, trucks, and deck up into the skater's feet and body.

This immersion can lead to improvised ways of inhabiting streets. What is being improvised here is an embodied, experiential and materialized form of urban knowing that emerges through repeated, incremental engagement with urban space – as Borden (2001: 193) put it, the 'skateboarder's senses' are 'historically produced'.

Improvisation is not, of course, a process that should necessarily be celebrated. It can, for instance in relation to water or sanitation provision, or in electricity or housing, represent an often desperate attempt to cope with severe hardship and poverty. Particular groups within impoverished settlements in Mumbai, for example, are forced to improvise sanitation facilities that constitute nothing short of a public health disaster for those that use them, as Figure 3 from Rafinagar, northeast Mumbai, suggests. If improvisation is forced upon many people in contexts of state welfare abandonment, complex coordinating systems often emerge as coping mechanisms. For example, reciprocal exchanges form the basis of insurance systems through which people borrow, lend, buy or sell between themselves. These improvised survival strategies include reciprocity in relation to a wide variety of services, from employment to water provision, and are usually mediated through close family and friends (e.g. Lomnitz 1977; Moser 1996; Gill 2000). Reciprocal systems are improvised, but emerge through incremental learning of urban change and negotiating networks of family and friends. They are used particularly in times of stress to cope with risks of, for instance, losing income or water provisions, and can serve both to reinforce and place stress on relations with family or friends, and to manage a field of uncertainties. These uncertainties range from unexpected drops in water supply due to contamination and damage to surface-lying water pipes, to complete removal and destruction by the state or landlords, forcing people to develop alternative systems of supply. For example, the removal of water pipes by the municipal state in Rafinagar in 2010 following a water 'shortage' that was blamed by élites on 'illegal slums', forced groups to utilize their personal networks and local knowledge to improvise reciprocal and, in many cases, far more expensive alternative private supplies. In the face of such acts of destruction of infrastructure and housing – acts that remove, deny or radically devastate urban dwelling and are all too common to many poor neighbourhoods in Mumbai – people draw on what they know to improvise alternative possibilities. In doing so, they draw on social infrastructure systems put in place over years that reflect an incremental learning of the sorts of vulnerabilities that cities might create and the kinds of coping mechanisms that people might put to work.

Reciprocity is learning through coordination that in turn emerges through dwelling. It addresses the shape of what Simone (2008b: 200) called

Figure 3 A makeshift toilet in Rafinagar, Mumbai (photograph used by permission of Renu Desai)

'everyday transactions' that 'facilitate, even at difficult and uncertain costs, the capacities of diverse urban residents to continuously make and adapt to conditions that keep the vast heterogeneities of urban life – its things, resources, spaces, infrastructures and peoples – in multiple intersections with each other'. Reciprocity is a multifaceted process through which marginalized groups learn as they adapt to, as Partha Chatterjee (2004: 40–41) put it, an 'uncertain terrain by making a large array of connections outside the group – with other groups in similar situations, with more privileged and influential groups, with government functionaries, perhaps with political parties and leaders' (and see Anjaria 2006 on street hawkers in Mumbai). These continuous efforts at urban maintenance and adaptation constitute porous learning assemblages, constantly drawing on and altering different urban knowledges. They are borne from the urgency to address marginality of different sorts – as Benjamin and Lacis (2004: 416–7) wrote

Figure 4 Painted trucks in Mumbai, used by permission of the author

in their famous essay on Naples, improvisation 'demands that space and opportunity be preserved at any price'.

Improvisation emerges through incremental urbanism. These examples show that improvisation is not straightforwardly spontaneous, 'of the moment', or mere *ad hocism*. It is learnt over time through coordination devices – i.e. the (re)construction of functional systems, such as those of reciprocity – that manage a range of different domains. Another quite different example of this is the large industry of subsistence painters and artists that work across urban India employed in vehicular art on trucks or busses – typically elaborate, colourful, and endowed with political, thoughtful and/or humorous messages (see, for example, Figure 4). As Swati Chattopadhyay (2009: 125) wrote of bus-art production in Kolkata, while these artists have their own distinctive signatures and flights of imagination, they draw on a rich repertoire of existing cultural resources and motifs, from 'three-dimensional representations of gods and goddesses, saints, and religious monuments such as the Golden Temple at Amritsar, Kali Temple at Tarakeswar, and the Kaaba festooned with decorations'. The improvisational quality of the urban artwork emerges through, on the one hand, artists borrowing both from a craft tradition (which is rural and urban) as well as a literary tradition of poetry, political sloganeering, and street talk customs, and, on the other hand, through infusing these 'with new images and events to create a realm of popular existentialism, advocating ways of negotiating and dwelling in modernity' (ibid. 129). Here, craft traditions that are made through learning-as-dwelling – i.e. embodied immersion in everyday practice and skill – are translated as they encounter different contemporary moments and instances (political debates of the day, moments in popular culture, etc.). This inseparable mixture of habits of craft and literature with popular images and slogans is obviously a means for making a living, but it is constituted by forms of knowing that are at once spiritual, popular, traditional, fantastical, and modern. Improvisation is, however, just one way in which people learn to negotiate cities through incremental experience. For example, given the diverse nature of many large cities, urbanites often need to learn to negotiate a wide range of groups, identities and places. In a multicultural city like Mumbai, this negotiation is a kind of 'worldly urbanism' that is learnt incrementally – although occasionally rapidly through sudden shocks and events – in conditions of radical uncertainty and hardship. To illustrate this, I will draw upon the example of street children in Mumbai.

Learning the Unknown City: Street Children in Mumbai

Mira Nair's 1988 film, *Salaam Bombay*, starkly illustrates the sometimes catastrophic predicament of uncertainty and learning that street children face. In the early part of the film, the main character, Krishna, a young boy, makes a speculative and surreptitious journey by train to Mumbai. He arrives to a crowded city of Bollywood action stars on billboards, imposing neo-Gothic architecture and heavy traffic. His first encounter is with a beggar: a desperate looking figure that sees the confusion on the face of Krishna as he stands and stares at the bustling street, a lonely figure out-of-place, and attempts to rob him. Krishna escapes and quickly runs into another street boy who steels his *paan* (betel leaf) and dismisses him as 'village trash'. Mumbai is cast here as a ruthless and unknowable metropolis, underlined by the contrast between provincial naïveté and a savvy, exploitative urban worldliness. Krishna's vulnerability is dramatized not just through his lack of money and protection, but his lack of knowledge of the city. The overriding message from these opening scenes is that Krishna's story is one of *getting to know* those parts of the city that will provide him with opportunities for survival and getting by. In 2006, I met with a group of street children in Mumbai that underlined this idea of the necessity of urban learning, expressed as a kind of urban worldliness. These children, all boys, live in one of the first sights many new domestic migrants encounter in the city, Mumbai Central railway station. What struck me was the crucial role that learning played in their everyday geographies of the city.

By 'worldliness' I am referring to the negotiation of different identities, spaces and temporalities that street children encounter in the city – a particular kind of subaltern street cosmopolitanism. This is not meant to romanticize the practices of street children, whose lives are subject to often unimaginable hardship and exploitation, but to identify how street children must learn to negotiate multiple groups and spaces in the everyday struggle to survive, through which they develop a sensitivity to the role that different individuals and groups play in urban life, as well as where and when to be in particular places. The modes of knowing the city for street children include pragmatism and fantasy, from access to employment opportunities, places for cheap food, places to sleep, and places to have fun or feed the imagination.

Like so many refugees to Mumbai, Rahul stepped off the Rajdhani Express from Delhi in 2002, at the age of nine, after he ran away from home following the death of his parents. When his train arrived into Mumbai Central, he had just enough money to buy some new clothes, and on his first night in the city he slept on a train platform. On his second day in the station he was approached by a local NGO support worker who patrols the station, and asked to come to a day centre across the road. He was given food and water and introduced to other boys, making him part of what has become for him a critical support network of friends and a communication network for employment opportunities. At the time of interview, he was selling magazines on trains and making between Rs 50 and Rs 60 per day. At 13 years old, he said he preferred Mumbai to Delhi because of the opportunities to get by.

The group at the day centre near the station is from all over the country, although most are from the northern states of Bihar and Uttar Pradesh or the more local confines of Greater Mumbai. Some of them arrived as young as seven, mostly from homes with domestic problems, of which alcoholism and violence are the most common. Some said that they enjoy the freedom of being without family in the city, and they are clearly a familiar bunch of friends – friendly abusive banter is constant between them, and they interact as close friends. The volunteers who run the centre said that sometimes the boys disappear for a few days, or even a few weeks, and have even been known to become lost by taking the wrong train or missing a stop on occasion. Some of the children retain links with home, and some even go back occasionally to deliver money. They talk of Mumbai as the best place to be in India, a place where money can be made. They have a wide range of ambitions, most of which include action-oriented exciting jobs, from becoming a soldier to joining the police. They say they are in Mumbai to get rich. In doing so, they are not afraid to commit the odd petty crime, although crime among most street children is rare. Some of them brag about robbing wealthier children of the same age, including of gold chains and gold rings that they subsequently pawn, and have even robbed other street children who enter the station.

Figure 5 'Hide-out' at Mumbai Central Railway Station, used by permission of the author

All of them (except one, a ragpicker) make their living from the trains or the station – selling goods on trains, polishing shoes at the station, and so on. When I asked Vilas what he would do without the trains, he held out his hand to indicate that he'd have to beg. They said the best trains to sell on are the busy intercities like the Rajdhani. The railway station is a shelter for them, particularly during the monsoon. They spend much of the rest of the time at the station, at their 'hide-out' – an open space between tracks with a ten-foot wall on one side, where they play and sleep (see Figure 5). But the station is also a threat. They are criminalized because of their presence in the station. Sometimes, a policeman who sits nearby chases them and, if he catches them, hits them with a stick. They have no permits to sell, and the police – often seeking bribes – use this as a pretext to chase or beat them. On one occasion, a minister was passing through the station and the police went on a mission to clear the space of street children. Eight of them were rounded up, taken to a nearby police station, and jailed. They got out as a group to use the toilet, and took the opportunity to run. On other occasions this has happened, people from the support centre have got them released.

Mumbai police have an added incentive, beyond cash bribes, in that they are expected to fill a monthly quota for getting street children off the streets and into remand homes. While some street children are grateful for the shelter during the monsoon, conditions in these homes are very poor and discipline is often severe. Many children view these places as prisons, notably the infamous remand home at Arthur Road jail. Although most of the boys are by now street-savvy and prone to ostentatious displays of confidence, many of them are terrified of the police. Vilas said that the police have beaten him by taking large sticks to his back, especially if they do not have Rs 10 or Rs 20 required to pacify them (one boy remarked that when he grows up he wants to be a policeman, so that he can beat those officers that currently beat him). For these children, the station and its railway tracks are an island experience that often brutally compresses the relationship between urban capital and marginalization. Like Krishna in *Salaam Bombay*, they never quite arrive – shelter and livelihood are always temporary, never settled. As Yiftachel (2009) remarks in relation to what he calls urban 'gray spaces', such marginalized lives 'are neither integrated nor eliminated, forming pseudo-permanent margins of today's urban regions'; they experience a 'permanent temporariness' (Yiftachel 2009: 89, 90).

Vilas, a little older and bigger than the rest, obviously carries some authority in the group. They are a very physical bunch, and he was not afraid to hit one of the others across the head or shout at them for playing the fool. He sells a range of items on the trains: stuffed into his polyester bag are toothbrushes, combs, pens, post-it notes, and much else. He got this from a trader in a settlement Malad, northwest Mumbai, who told him he must return Rs 1000 from the bag and that he could keep the rest. He aims, he said, to become an encounter specialist (infamous Mumbai police-assassins involved in fighting gangs). Vilas is distinguished from the others partly because he is older and bigger, and partly because he is more 'street-wise', i.e. because of what he knows of the city in comparison to the others; he has learnt of good places to go and people to speak to for finding work. He possesses, in short, knowledge that is intersubjectively valuable.

While many of these children, like many others of the hundreds of thousands of children living on the city's streets, did not necessarily set out to come to this city, Mumbai is for them a centre of opportunity and heroes in the shape of Bollywood action stars and police encounter specialists, a 'charismatic city' in Hansen and Verkaaik's (2009) terms (and see Prakash 2006, 2010; McFarlane 2008b). The worldliness that they live is produced through incremental learning, in groups, ways and means to negotiate the uncertainties of multiple marginalization. Often working together, they draw upon their experiences and the knowledge that the different children within the group have of the city to tap in to employment opportunities, places for cheap food, places to sleep, places to have fun, and when the best times of day are to pursue these activities. Getting lost or attacked reveals the uncertainty and vulnerability integral to this urban learning. They do not negotiate different spaces, groups and identities in the city with knowledgeable ease, but through participation in social groups where they learn a 'feel' for different city geographies – in the words of Hansen and Verkaaik (2009: 8), writing in a different context, they are 'able to convert the opacity, impenetrability, historicity and latent possibilities of urban life into a resource in their own self-making'.

That they learn most of this incrementally, through gradually gaining a sense of how things work over time, does not mean that they do not also learn through sudden events. Indeed, they learn through sometimes terrifying events, like being arrested or being beaten by the station police officer, but much of what they learn over time is about how to negotiate relations between each other and external individuals and groups like the non-governmental organization (NGO), the informal employers, the train customers, and the police, as well as learning the rhythms of when is a good time to use the 'hide-out' or to go around the station unseen by the railway police. None the less, it is their ability to learn this kind of worldliness – a particular kind of dwelling as education of attention – that dramatizes and addresses the marginality of these children. These forms of learning are dependent not on knowing the city as if it were a field of informatics, but on learning the city through the senses in relation to fear, hope, fantasy, fun, wonder, and so on: through a haptic incremental immersion rather than simply a cognitive or optical view, as well as through the shifting power relations and gestures of group dynamics.

These children have, by necessity, what Elijah Anderson (1990) called in his study of street life in a low-income African-American neighbourhood, 'street wisdom' (and see Bridge 2005: 51–4). As Gary Bridge showed, street wisdom is achieved in a series of stages. There is 'mental note-taking' of people in the street (or train station) and their activities at different points in the day, which through repeated interactions and 'note-taking' leads to a particular way of knowing through what Anderson called a kind of 'field research' (Bridge 2005: 52). If such street wisdom helps these children to cope or escape hardship, it rarely gives them much of an upper hand, and in their negotiation with strangers they are rarely able to exhibit the sort of confident street wisdom that more privileged urbanites might deploy. Learning through incremental urbanism calls forth the importance of both space and time as different groups seek to cope with, improvise and negotiate the city through an often uncertain array of different actors, power relations, and inequalities. To dwell in the city is not just to move through space but through the negotiation of multiple actors and temporalities. The next section examines the temporalities of everyday urban learning by considering the notion of *rhythm* in more detail.

Learning, Rhythm, Space

What does this discussion of learning through dwelling bring to conceptions of urban space? What is immediately clear from the preceding discussion is that urban space must be first and foremost understood as a relational constitution, i.e. urban space is produced by assemblages – sociomaterial alignments, sometimes stable, sometimes precarious – that make up the continuities and discontinuities of urban dwelling. These assemblages take various spatial forms and functions, of which three in particular stand out. Firstly, assemblages can structure the possibilities of dwelling. They are not only produced by dwelling practices, but curtail the possibilities of dwelling, meaning they can be enabling or disruptive. Housing or infrastructure materials within informal settlements, for instance, are central to how many people learn to cope with and negotiate urban life and poverty. Secondly, assemblages can, none the less, exceed these structural confines. This excess takes at least two forms: as both to given fields of possibility, and to that which is immediately present: they can generate random juxtapositions, they can be imagined but not realized, and they can be seen, felt, heard, and non-cognitive. Thirdly, assemblages have no necessary temporal or spatial template: they can be fleeting, lasting, instantaneous or laboured, and they may be near or distant, present or absent.

There is, of course, a long history of urban thinking that seeks to capture the liveliness of space that these three issues point to. We might think for instance of the dizzying, jarring and noisy scenes of movement that fill Walter Benjamin's vivid descriptions of cities like Berlin, Naples or Paris (e.g. Bullock and Jennings 2004), or Henri Lefebvre's (2004) account of the generative and relentlessly exploitative nature of space created through capitalist production and accumulation, and his discussions of the passion and action of everyday life. Or the politics of Guy Debord's anti-spectacle everyday urbanism (e.g. McDonough 2009), and Manuel Castell's (1983) early interventions on the possibilities of grassroots movements to reshape urban spaces. Outside of the 'Western urban canon' there is an equally rich history of accounts of everyday urban spatialities. Edgar Pieterse (2008: 113) for instance, examined work that traces the 'below-the-radar sets of small actions to gain a little piece of pavement, or a few square foot of floor space, or illegally tapped kilowatts from a government-owned power line, or enough invisibility to duplicate copyrighted goods for sale at informal markets' (and see the important work of Simone 2004a,b, 2008a,b; Bayat 1997). In these different accounts, urbanism – to paraphrase Ingold (2008: 1808) – does not so much exist as *occurs*; it is a sociomaterial achievement continuously remade through different encounters, labours and mobilizations. These accounts focus on the liveliness of urban dwelling as a process of forced improvisation in contexts of often profound urban inequality, and they draw attention to the multiple rhythms that constitute the urban experience.

In my own research in Mumbai, I am often struck by the difficulty of describing the spatial discontinuities of the city's eventful and disruptive everyday encounters, the changing daily and nightly rhythms of recurring practices and random thrown-togetherness, the juxtaposition of cacophonies of noise, exhaustion and peace that characterize so many of its neighbourhoods, and the atmosphere of sociability that is generated in the city's streets on often sedate evenings – in short, some of the 'key qualities of a city-type ambience, of "citiness"' (Healey 2002: 1780). Indeed, if dwelling is an education of attention – an attunement of perception through the urban environment – it is through combinations of tactile, sensual, personal and explicit knowledges that we come to learn and relearn the city through dwelling – what Ingold and Kurtilla (2000: 189) refer to as a 'multisensory awareness of the environment' central to 'spatial orientation and the coordination of activity' (see Crang 2001; Middleton 2009). By way of example, I will highlight one snapshot from extended fieldwork in Mumbai in 2007. The purpose of doing so is to open up a wider discussion around rhythm that reveals a series of mundane ways in which people learn the city through dwelling.

At the time, I was researching in Bandra, west Mumbai. Nearby where I was staying there was a paan shop run by a young man named Aza. Paan is a betel leaf that is filled, usually with lime paste and areca nut, and chewed as a digestant. Aza's paan shop, set on a corner just off Bandra's busy and fashionable thoroughfare, Linking Road, does a steady business at night, especially after dinner, partly due to the association of paan with digestive qualities. He has had to learn to use an extremely small space optimally. His wooden hut is less than two metres by two metres, and for the privilege he has to pay Rs 20 lakhs (not far off £30,000 per year) to the municipal corporation – 'Bombay is the most expensive city for rents in the world', he complains. Aza works at the micro-level. He has just enough room to stand. On his right is a waist-high fridge filled with soft drinks. Above the fridge, on the shelves against the back of the hut and on the wall to his right, cigarette boxes are crammed in. In front of him there is a waist-high table upon which is stacked silver tins containing his paan ingredients – herbs and spices – and a stack of two different sizes of betel leaf. Bottles of water are squeezed into the space below. On a shelf to his left are two phones, one for local calls and the other for international calls. Like the urban house sculptor in Paraisópolis, but through quite a different form of craft, Aza's hut is an assemblage learnt through haptic immersion in a confined space.

Every half a minute or so a new customer appears requesting his particular mix of paan wrapped in betel leaf. Friendly conversation sometimes ensues. 'You have to have the talk', Aza tells me, 'speech – you have to know how to relate to your customers so that they will come back.' With some of the regular local customers he has built up a rapport over the five years in which he has been based on this spot. The possibility of this rapport is based in no small part on the mood and atmosphere of these warm evenings, when people slow down, take their time to chat, and inhabit the street in a markedly different way from during the day. In the street scene of which Aza's hut was a part, people sauntered in a relaxed manner, negotiating the parked cars and enjoying the gentle bustle of people strolling or buying chai and paan set against the dim street lights illuminating large evergreens. This is not to say that Aza's customers only take the time to talk with him during sedate evening strolls, but that the particular atmosphere of relaxed saunter clearly meant that more people were inclined to take time to talk with Aza, no doubt facilitated by his always smartly dressed appearance and charismatic personality. But whatever his natural confidence, Aza has had to learn his interactions with his customers over time. In addition, the particular rhythm and pace of the evening, particularly that of the slow relaxed stroll, matters for Aza's custom.

Walking is central to how we come to learn urban space. This is not just a spatial learning of urbanism, but a temporal practice of place, and is part of a wider set of rhythms that characterize urban places (Wunderlich 2008; Middleton 2009). As Filipa Wunderlich (2008) argued in her phenomenology of urban walking, walking is an unconscious activity through which we firm up and alter our relationship to urban places. Walking is a key form of learning through dwelling, a practice through which we become 'immersed in temporal continuums of social everyday life activities fused with spatial and natural rhythmical events' (ibid. 26). Walking is a pre-reflective form of knowledge, although it sometimes entails discovering and transforming our conception of urban places through the making and remaking of connections to past, present and future, between the real and imagined, and through noise, smell, vision and touch (De Certeau 1984; Seamon 2000; Rendell 2006; Bridge 2005, on walking as 'strategic rationality'). Motion creates an experiential sense of direction, perspective, distinction, and lay-out in the city. As a form of 'circumambulatory knowing' (Ingold 2004: 331), walking is a particular kind of education of attention to the urban environment through which people experience and learn urbanism, an embodied condition of thinking and perception in constant flow (Winkler 2002; Wunderlich 2008: 128). There are, of course, different forms of walking, and these can entail distinct forms of urban learning. Here I highlight 'purposeful walking', 'alert reverie', 'discursive walking' and 'conceptual walking'.

Walking can also be experienced as a form of urban disengagement. For example, in what Wunderlich (2008: 131) called 'purposeful walking', for instance of people walking from train stations to work in the morning, walking might be experienced largely through listening to an iPod, talking on a mobile phone, or eating a quick, late breakfast. There is a spatio-temporality to this disengagement, as Lefebvre (2004: 28) wrote in his discussion of rhythm and everyday life: 'He who walks down the street, over there, is immersed in the multiplicity of noises, murmurs, rhythms (including those of the body, but does he pay attention, except at the moment of crossing the street, when he has to calculate roughly the number of his steps?).' This fluctuating relation between attentiveness to the urban environment and individual disengagement – what Iain Sinclair (1997) called an 'alert reverie' – is characteristic of purposeful drifting, and can create an imaginative geography of embodied urban connections that align multiple spaces and times:

> Alignments of telephone kiosks, maps made from moss on the slopes of Victorian sepulchres, collections of prostitutes' cards, torn and defaced promotional bills for cancelled events at York Hall, visits to the homes of dead writers, bronze casts on war memorials, plaster dogs, beer mats, concentrations of used condoms, the crystalline patterns of glass shards surrounding an imploded BMW quarter-light window. ... Walking, moving across a retreating townscape, stitches it altogether: the illicit cocktail of bodily exhaustion and a raging carbon monoxide high. (Sinclair 1997: 4, cited in Borden *et al.* 2001: 19)

The city that is 'stitched together' through walking is a relational city of multiple times and spaces, memories and bodily experience, and constituted through an assembly of translocal commodity chains, ecologies, histories and unfolding events.

If walking appears as an individual experience, it is of course an experience that is socially mediated. As a form of dwelling, walking entails learning how to perform socially 'as part of temporal patterns of economic and social life, performed at different paces and rhythms, tracing our paths and intersecting other people's life routines' (Wunderlich 2008: 130). For example, what Jane Rendell (2003: 230) called 'urban roaming', or what Filipa Wunderlich (*ibid.* 132) called, after De Certeau (1984), 'discursive walking', is a form of walking specifically geared around sensorially experiencing the surrounding urban environment, mediated by social relations such as those of class, race, ethnicity, gender, and social norms around the nature and behaviour of 'acceptable' bodies – walking is an individual performance, but it is also social reproduction (e.g. see Munt, 2001, on the lesbian *flâneur*). Here, Benjamin's (1999) *flâneur* experiences the urban landscape and its flows and rhythms in ways that are socially, politically, economically and ecologically mediated.

But walking can be, within the expectations of each urban situation, a form of urban improvisation (Bridge 2005: 51; De Certeau 1984). Closely related to this is Wunderlich's (2008: 132) notion of 'conceptual walking', or walking in a reflective mode – a process of becoming acquainted with particular urban places: 'It is used consciously as a way to get to know the city and uncovers features not usually noticed in our everyday life.' Walking provides the possibility to experience the city differently, and can create a sense of liberation from power as the walker carves out a city that differs from the bureaucratic plan; through walking, people can learn to dwell in a more personalized city. For example, one important urban tradition of conceptual walking is the Surrealist and Situationist *dérive*, or drift, a process of rethinking urbanism by rendering it strange. For Debord (1956, reprinted in McDonough 2009: 78), *dérive* was 'a technique of swift passage through varied environments', an attempt to unsettle the city's bordered, administered, mapped and controlled 'psychogeographic relief' and to work towards a more creative, flexible form of urban living that enabled greater freedom of movement, design and inhabitation. Its aim is both to expose and render the psychogeographic city of restriction – to reveal, as Lefebvre put it, 'the growing fragmentation of the city' (Merrifield 2002: 97) – and to experience urbanism alternatively as a disorientating, bewildering and unknown city that might be relearnt – 'one day, cities will be built for *dérive*' (Debord 1956, in McDonough 2009: 85). As Steve Pile (2001: 264) wrote in a different context: 'Exploring the unknown city is a political act: a way of bringing to urban dwellers new resources for remapping the city.'

These and other walking practices constitute, along with other bodily, social, economic, political and ecological urban rhythms, the polyrhythmical nature of urbanism (Lefebvre 2004). We come to know – implicitly and explicitly – these rhythms through social norms, practice, experience, memory, and haptic recollections that allow us to apprehend and negotiate the movement of diverse rhythms, and which allow us to dwell in the urban present: 'to *live* it in all its diversity' (Lefebvre 2004: 36). For Lefebvre (2004), these different rhythms, with their specific temporalities, speeds, cycles, forms of repetition and difference, sensory atmospheres, geographies, presences and absences, cumulatively provide a glimpse of urban order: an order created by, for instance, the state (present even if visually absent, e.g. through law, regulation and investment), the divisions of labour which produce work and leisure time, the social norms that dictate routines such as school hours or particular forms of interaction between strangers, and the lay-out of streets, neighbourhoods and services. If Lefebvre chooses the window as the medium through which to gaze upon urban street rhythms in his famous 'Seen from the Window' chapter in *Elements of Rhythmanalysis*, the window is also a practical metaphor – an urban learning device – for coming to know something of the unknowable city: 'Opacity and horizons, obstacles and perspectives imbricate one another to the point of allowing the Unknown, the giant city, to be glimpsed or guessed at, with its diverse spaces affected by diverse times: rhythms' (Lefebvre 2004: 33).

Assemblage thinking is alert to the temporal dynamism of changing urban rhythms, where the urban street, for instance, is a place of changing density and quietude, altering functions (shopping, protest, sociality, passage, etc.), growth and death, changing aesthetics, and the 'flows of life, traffic, goods and money that give the street its intensity and sense of place' (Dovey 2010: 16). As Nicholas Tampio (2009: 394) has put it, 'the brilliance of the concept of assemblages is that it describes an entity that has both consistency and fuzzy borders. ... [It] has some coherence in what it says and what it does, but it continually dissolves and morphs into something new.' The challenge here is to come to terms with the spatiality of urban learning assemblages as a constant relational co-production in which the possibilities for dwelling the city are altered, generated and delimited across multiple rhythms. Importantly, rather than figuring the multiple space-times of urban learning as localized and bounded, assemblage brings to a conception of dwelling an emphasis on the processual and mobile. People learn to negotiate changing and multiple urban rhythms through dwelling, but assemblage serves as a reminder of the need to unsettle the heavy connotations of localism and rootedness that the concept of dwelling has traditionally carried. Indeed, Ingold (2008) has come to regret his emphasis on dwelling, preferring now to speak of *inhabiting*. Citing Steve Hinchliffe's (2003) concerns with the notion of dwelling, Ingold acknowledges that despite his effort to transcend a notion of dwelling as a kind of territorial and earthly romanticism, 'it is, nevertheless, true that the concept of dwelling carries a heavy connotation of snug, well-wrapped localism' (Ingold 2008: 1808). However, rather than dismiss the notion of dwelling, another route is to consider how we might rethink dwelling *through* the mobility and translocality of urban rhythm.

Cities have long been transit points of mobile exchange that have propelled multiple rhythms of different temporalities and speeds, from daily rhythms of work, school and nightlife to economic cycles of accumulation and disinvestment and cultural flows of migrants. Assemblage offers a spatial image of cities as incubators of mobile socialities, where the reterritorialization of the city is predicated on multiple durations of spatiotemporal displacement. The city emerges here as a historically relational composite of cohabiting spatiotemporalities where the condition of possibility of dwelling rests in part upon this very discrepant mobility. For example, as Venn (2006b: 47) wrote of the material geographies of diasporic groups within cities, the unfolding of urban dwelling emerges from long histories of pluralized mobility:

> ... evidenced in the everyday in the decor and style of houses and buildings, or indeed in graffiti, and, generally, in the translation of place through their reiteration (their repetition with-and-in difference) in a new spatiality – say, China Town in Los Angeles, or Little India in Singapore, New England in New York ... a form of memorialisation and socialisation of living space.

Far from conceiving this constitutive mobility of dwelling as a problem, Venn's account stressed the advantage of the assemblage concept for envisaging 'disequilibrium' as normal and of dwelling in the city as an ongoing spatiality of changing rhythms of different temporal lengths that people learn to negotiate (even if that involves learning *not* to negotiate them, i.e. to avoid, ignore or retreat from them). Learning through dwelling is a relational process produced not through an internal, bounded experience, but through the multiple rhythms, temporalities and spatialities that shape everyday urbanism. In this context, incrementalism emerges as one of the multiple rhythms that shape urban learning, albeit a particularly important one. But learning, as Chapter 1 argued and as has been reiterated at various points in this chapter, is not reducible to dwelling and must be understood in a broader matrix alongside translation and coordination. Indeed, learning can emerge through translation and coordination to operate as a tactic that offers very different possibilities for dwelling, for instance in the learning tactics of urban social movements. In the final part of the chapter, I consider how learning can operate as a tactic that explicitly seeks to address urban marginality. I offer a conceptualization of 'tactical learning' to name the process whereby learning can open new possibilities for urban dwelling.

Tactical Learning

Perhaps the most influential work in debates on tactics is that of Michel De Certeau (1984; De Certeau et al. 1998) on strategy and tactics. Strategy refers to the 'calculus of force relationships' sustained by a base of power once a subject becomes isolated (e.g. a governmental institution) (De Certeau 1984: xix, 35–6). Strategy postulates a place that can serve as a delimited base and has independence in relation to its circumstances and others. A tactic is a fragment that manipulates events and turns them into opportunities – its operation lies in 'seizing the moment', hence De Certeau's (1984: 39) emphasis on the 'utilization of time'. A tactic is a 'calculated action determined by the absence of a proper locus'; unlike strategy, it does not demarcate an exteriority necessary for its autonomy: 'The space of the tactic is the space of the other' (De Certeau 1984: 37). Tactics 'traverse' and 'infiltrate' systems by playing out 'the guileful ruses of different interests and desires': 'It must vigilantly make use of the cracks that particular conjunctions open in the surveillance of the proprietary powers. It poaches them. It creates surprises in them. It can be where it is least expected' (ibid. 34, 37). Tactics refer to the kinds of action that are possible once people have been marginalized by different strategies, and include a range of everyday forms such as speaking, walking, reading and shopping. De Certeau is concerned with the ways in which, in trying to get by in ordinary life, people use practical knowledge of how things work that can then be translated into different uses and contexts. His notion of tactics, then, is rooted in learning through dwelling.

In the conception of strategy and tactic, there is a risk of implying too rigid a separation of the official and the everyday. Related to this, there is a latent potential in De Certeau's work of romanticizing marginality (Ruddick, 1996; Iveson, 2007). For instance, he has surprisingly little to say conceptually about the spaces between compliance and resistance (Napolitano and Pratten 2007; Hansen and Verkaaik 2009). In this latter respect, James

Scott's (1985) contemporaneous work on 'weapons of the weak' in relation to everyday forms of peasant resistance is particularly useful. Scott's (1985: 29) 'weapons of the weak' referred to 'foot-dragging, dissimulation, false compliance, pilfering, feigned ignorance, slander, arson, sabotage, and so forth'. If this overlaps with the idea of tactics in that it is based primarily on forms of learning through dwelling, Scott was none the less concerned with more visible forms of resistance. As with tactics, these weapons often – although not exclusively – emerge through everyday forms of knowing: '[These forms] require little or no coordination or planning; often represent a form of individual self-help; and they typically avoid any direct symbolic confrontation with authority or elite norms' (ibid.). One example he gave is 'the quiet, piecemeal process by which peasant squatters have often encroached on plantation and state forest land' (Scott 1985: 32). These forms are 'often covert, and concerned largely with immediate, de facto gains' – forms that Scott explicitly referred to as 'informal', in contrast to institutionalized politics that are 'formal, overt, concerned with systemic, de jure change' (ibid. 33). He elaborated on this throughout the book in references to 'implicit understandings' and 'subcultures of resistance' that are known but not always made explicit. Later, in his (1990: xii) *Domination and the Arts of Resistance: Hidden Transcripts*, Scott called these multifarious processes the 'infrapolitics of the powerless'.

But Scott was also concerned with the ways in which resistance does not just emerge from everyday dwelling, but can stand apart from and intervene in dwelling. While forms of resistance that emerge through, for example, incremental urban learning such as house construction and walking are important, I want to consider here forms of urban learning that do not emerge from dwelling specifically but which none the less serve to open the possibilities of dwelling. Tactical learning can arise from incremental learning (as we will see in Chapter 3 in relation to informal housing), but it can also arise outwith forms of urban dwelling. For Borden et al. (2001: 13), for example, tactics are less about everyday dwelling and take the form of more exceptional interventions, including 'attempts to solve urban problems with housing programs, planning policies, and political agendas; and they may also be attempts to reconceptualise the relation between the city and the self. ... Tactics aim to make a difference.' But how might we understand this infrapolitics of knowing that disrupts urban dwelling? It is in this context that I seek to extend De Certeau and Scott to consider how forms of tactical learning intervene in and open up specific possibilities of dwelling.

One useful reference point here is Hansen and Verkaaik's (2009: 13) notion of 'urban infrapower': durable assemblages of resources and connectivity. Urban infrapower names three domains through which people interpret and act upon the city. The first is *'sensing* the city', 'i.e. reading, reproducing and domesticating the urban soundscapes, the visual overflow, the styles, smells and a physical landscape that can be read through everyday mythologies of past actions, heroes, martyrs, events, danger' (Hansen and Verkaaik 2009: 12). Here, sensing maps on to the notion of dwelling: we learn sensorially, through immersion in the urban landscape. The second is *knowing* the city', 'in the sense of decoding it, managing its opaque and dangerous sides, controlling and governing the urban landscape' (ibid. 12–13). Note here that they are not referring to any particular group or territory; we might be talking about policy-makers, community groups, street hawkers or taxi drivers. This includes everyday incremental urban knowing that enables particular forms of infrapower, for example in relation to poor neighbourhoods: 'Although popular neighbourhoods do appear to resist legibility, in James Scott's (1988) sense as a gaze of the state, such spaces are nonetheless navigated and interpreted by their residents on a daily basis' (Hansen and Verkaaik 2009: 15). These neighbourhoods are home to people who possess 'superior knowledge of these densely populated spaces: the hustler, the hard man, the wheeler-dealer' (ibid.). This points to everyday modes of knowing the city, to the 'performative competence' of urban registers, and to the 'urban specialist' 'who by virtue of their reputation, skills and imputed connections provide services, connectivity and knowledge to ordinary dwellers in slums and popular neighbourhoods' (Hansen and Verkaaik 2009: 16; Ferguson 1999). In the schema of learning as translation–coordination–dwelling, Hansen and Verkaaik's notion of 'knowing' refers to coordination devices – for example, the 'wheeler-dealer' is someone able to bring together a range of different knowledges and contacts and put that resource to work in various ways. The third domain is the capacity for 'urban gestures' and actions – registers of public performance that are known to people in specific neighbourhoods, whether as individuals or as crowds (ibid. 13) – in other words, the role of gesture to facilitate learning by translating meaning between different groups.

As useful as this schema of urban infrapower as *sensing, knowing* and *gesturing* undoubtedly is, for my purposes I would not wish to separate out 'knowing' from 'sensing' and 'gesturing'. Rather, sensing and gesture are important ways in which knowing takes shape and is communicated. In Hansen and Verkaaik's account, knowledge is positioned as neither 'officially codified' nor 'concealed or secret', but instead as a form of 'brokerage' or tactical purpose (Hansen and Verkaaik 2009: 20). Infrapower is always emergent because it only shows itself in action or outcome, and is reinvented through action. The forms of learning that constitute infrapower only become forms of power when they enable opportunities – i.e. become realized as tactics. The notion of infrapower is useful, then, in that it highlights a caveat in De Certeau's tactics and Scott's weapons of the weak in that it draws attention to the fact that tactics and weapons must be *learnt*. As Hansen and Verkaaik indicated in passing at several points, those mediators who possess urban infrapower because they are 'in the know' need to have learnt these particular modes of knowing the city. But despite the useful emphasis here on infrapower as an emergent property of learning that has the capacity to intervene in dwelling, I want to push this discussion further to consider how learning can operate as a tactical form of infrapower that can radically disrupt everyday dwelling. By way of example, I will draw upon fieldwork conducted on an urban social movement in Mumbai, the Federation of Tenants Association. In this example, tactical learning emerges not through everyday urbanism, but through an intervention by a social movement that depends upon learning through translation and coordination and that opens the space of possibility for urban dwelling.

The legal assemblage

The Federation of Tenants Association, a male-dominated organization led by predominantly middle-class housing activists, works to raise awareness of existing laws and regulations amongst marginalized settlements that protect housing rights. As one of the Federation leaders, Deshpandy, claimed in interview, the state has little interest in publicizing these regulations because it threatens the 'state-corporate development nexus', so the Federation attempts its own publicity through large public meetings within popular settlements throughout the city. One site in which the Federation works is Bharat Nagar, a predominantly Muslim settlement in central Mumbai. This old neighbourhood is under threat from predatory developers due to exponential increases in real estate costs throughout the last 15 years caused by the development of the nearby Bandra-Kurla financial and administrative complex. There have been accusations made by community members that the developers have used bribes and threats of demolition in order to get people to move from the land. The act of demolition is common practice within many popular settlements in the city that find themselves occupying land that has become lucrative. The often unpredictable demolition that many settlements live with constitutes the transformation of space by the state for (and often with) capital – in Wacquant's (2008) terms, the militarization of marginality by the neoliberal punitive state. It reflects the investment of the state in particular populations over others, an unequal biopolitical investment reminiscent of Ong's (1999) conception of the 'postdevelopment state' marked by a graduated sovereignty that divides the population into different mixes of 'disciplinary, caring and punitive technologies' (Ong 1999: 217).

In the spring of 2006, the Federation were in communication with local leaders in the area and organized a mass meeting that sought to reach local people at large. That the meeting went ahead at all was remarkable: it was alleged by the Federation that the developer in question had tried to prevent the meeting by putting up posters stating that the meeting had been cancelled, then when word got around about the posters being false they put up new posters saying that the venue had changed. All of this ironically served to raise the profile event by raising interest. The federation leaders, along with other representatives, each took to the stage to address the crowd, and a question-and-answer session took place after the speeches, lasting over an hour. The objective was to inform people about, firstly, the exact worth of the land given the steep rise in land prices in the area, and, secondly, the rights they had under Mumbai's controversial Slum Rehabilitation Authority (SRA) scheme to act as a collective developer themselves by forming cooperatives (see Chapter 3).

The Federation's work has shown that many people in the city living in informal settlements simply have no knowledge of formal regulations: it is a movement predicated specifically upon a politics of learning. Learning here dramatizes the nature of marginality as a constitutive outside: many of the tenants of this settlement live illegally and this is, of course, productive of their exclusion from security and services, yet some aspire in this instance to know the law in order to provide some access to a world of security and services. Around 3000 people came to the meeting itself, almost everyone in the settlement. The meeting was sharply divided along gender lines, both in terms of the male leaders who confidently strutted the stage and the segregated audience that channelled the women to one side of the meeting space. The atmosphere was highly charged, especially during the long and animated speech by the Federation's charismatic leader, Chandrashekhar Prabhu. Prabhu is a natural performer, and delivered an expressive, stylish, dramatic and at times crowd-rousing speech. As the meeting went on, more and more people were enthused by what they were hearing from Prabhu about collective 'self-development'. Prabhu explicitly tapped into a Ghandist tradition of community-driven anti-establishment resistance.

Speakers encouraged people to form cooperative housing societies, pointing out that this would allow them to develop the land themselves and exclude the builders. Federation leaders recounted stories of successful self-development schemes in other parts of Mumbai. One NGO leader traced the various steps that local cooperatives might embark upon: sorting your own accounts, organizing yourself without builders, dealing with contractors and payments on your own, and so on – here, tactical learning emerges not through dwelling, but as a chain of translation between legal and non-legal knowledges, with the notion of the 'cooperative' operating as a coordinating device. This 'self-development scheme' would involve cooperatives managing construction and selling some of the housing themselves and using the profit (after construction costs) to fund a cooperative bank account that would be used to manage building maintenance and pay for property tax. The tenants themselves would pay nothing. A few days later, the situation in Bharat Nagar became volatile, with some in the area grouping with the developer and some opting for the self-development cooperatives. Neighbours and families were quarrelling. The details and regulations themselves have caused anxiety, with people struggling to understand Mumbai's complicated development system of floor space indices and transferable development rights, and for some accepting quick bribes can be simpler and faster. If the Federation's intervention provided a learning opportunity for rethinking the possibilities of urban dwelling, some were left daunted by the uncertainties of understanding legal information and the complications of forming cooperatives. In the weeks that followed, Federation leaders began smaller-scale meetings by splitting the area into small plots of a couple of hundred houses. These meetings were designed to help to answer queries, and to try to persuade people to form plot-based societies, to function as coordination devices for tactical learning. The future of the settlement remains uncertain.

As tactical learning, the Federation's work addresses urban marginality through the domain of the law, with the aim of extending the formal security that mainstream residential groups in the city experience to a neighbourhood with precarious rights. This learning is facilitated by key charismatic individuals like Prabhu, who amongst this audience performs to impressive effect a distinct regime of urban infrapower (sensing, knowledge, gesture). The Federation is using formal, codified legal information, but – and this is crucial to the success of the movement – communicating it through everyday, informal registers. The charismatic figure of Prabhu – who not only knows the formal laws of the city, but also knows the affective realm of gestures and speech that are necessary to put the law to work – demonstrates the near-impossibility of prising the formal and informal apart. In addition, if at the centre of this movement is learning about rights and procedures, that learning is not simply a process of encountering codified knowledge. Rather, the charismatic performance of Prabhu facilitates learning through the coordination of different domains, including the powerful role of gesture and speech in Prabhu's public performance, the promotion of discourses of 'self-development', people's capacities for patience in a hostile, fraught environment of threats from developers, the invocation of histories of Indian resistance, the desire for security of tenure, and of sharing stories of successful examples in collective tenure from other parts of the city. Learning through the Federation's work in Bharat Nagar is an assemblage of multiple histories, knowledges, sites, materials and groups, as well as of emergent solidarities, fears, hopes and uncertainties. It is an urban tactic of resistance that aims to advance the prospects of the urban marginalized, and which emerges through the translation of codified, legal knowledge through forms of sensing and gesture.

Conclusion

Incremental learning is an experiential geography of urban accretion. It takes a diversity of forms, from the bit-by-bit ways in which informal housing is added to or altered to meet new needs or possibilities, to pooling contacts and resources to develop new economic opportunities or building reciprocal exchange systems amongst family and friends. Incrementalism is a central process of urban dwelling, an attunement of perception to what urbanism – in conditions of often extreme inequality – might enable and delimit, and to how people might negotiate it. As a form of urban dwelling, incremental learning names the everyday practicalities that constitute learning assemblages, and the perceptual fields that are both shaped by and in turn alter those practical enactments. Incremental learning can lead to improvisation, positioning improvisation not necessarily as a sudden moment of inspiration but as a form of tweaking and altering urban arrangements over time and space, for instance through relations of reciprocity. But, if dwelling connotes a bounded localism, assemblage offers a spatial corrective through an openness to how relations are gathered together, aligned, or transformed through urban learning. As the spatiality of dwelling, the focus on rhythm and mobility that assemblage brings disrupts the bounded, grounded baggage that dwelling connotes, opening a wider imaginary of urban spatial topology.

Urban learning in the context of improvisation involves acting within assemblages of multiple relations, between family and friends, sustained explicit infrastructural arrangements and systems of borrowing and lending, anticipating the rhythms of the city over day–night, week and season, the legal and illegal, the modern and traditional, the new and the habitual. In the case of the street children at Mumbai Central, their ability to learn a kind of worldliness – a particular kind of education of attention – reveals urban learning not as a field of informatics but as getting a 'feel' for dwelling the city through the senses in relation to fear, hope, fantasy, fun and wonder, as well as through the gestures of group dynamic. If incremental learning connotes a gradual temporality of urban change, it is important to reflect upon the multiple spatiotemporal rhythms that inform urban learning experiences. Living urban lives involves learning how to participate in and negotiate urban rhythms: rhythms that both shape and are shaped by differential capacities to explore and sift and sort urbanisms in contexts of often profoundly unequal resources and power relations. People learn to negotiate the polyrhythmic nature of urbanism through practice, experience and memory, for instance through different forms of walking in the city.

In conceptualizing the ways in which different groups address urban marginality, De Certeau's (1984) tactics and Scott's (1985) weapons of the weak are useful starting points, particularly when brought into dialogue with Hansen and Verkaaik's (2009) account of urban infrapower as sensing, knowing and gesture. Infrapower draws attention to the fact that tactics and weapons must be *learnt*. As Hansen and Verkaaik suggest, those mediators who possess urban infrapower because they are 'in the know' (such as Prabhu) need to have learnt these particular modes of knowing the city. Tactical learning can emerge through everyday dwelling, but it can also emerge through translation (e.g. of legal knowledge) and coordination (e.g. of cooperatives, or through figures like Prabhu) that open alternative possibilities for urban dwelling. Tactical learning is a variegated set of resources that responds to what Hansen and Verkaaik (2009) call the city's 'constitutive unknowability' by performing a crucial role in how people resist in the city in often extremely difficult circumstances. The different forms of urban learning discussed in the chapter constitute, to borrow from Chattopadhyay (2009: 135), 'an assemblage of fragmentary elements in space through which subaltern groups make room for themselves within a spatial structure that is not conducive to their existence'. The next chapter continues this theme by critically examining the tactical learning of an influential urban social movement, *Slum/Shack Dwellers International*. In doing so, it examines how a focus on learning can reveal important practices and politics that constitute social movements.

Chapter Three

Learning Social Movements: Tactics, Urbanism and Politics

Introduction

In the summer of 2008, I interviewed the leader of a small non-governmental organization in central São Paulo about her organization's work and connections to an international urban movement of 'slum' activists. Anaclaudia Rossbach, the Director of *Internacao*, explained the importance of being part of this international movement: 'We [*Internacao*] survive because of their methodologies.' For Anaclaudia, being part of this wider international movement was first and foremost about *learning* from other people's experiences of addressing urban poverty. The movement she referred to, Slum/Shack Dwellers International (SDI), is an international movement working with urban poverty in slum settlements in over 20 countries, and Brazil is one of the more recent recruits in the shape of *Internacao*. I asked Anaclaudia about her experience of connecting to SDI, and she replied: 'Talking is important, but I guess seeing it on the ground … you see that it has credibility.' Anaclaudia first met members of SDI through a meeting of Cities Alliance, a network of governmental and non-governmental organizations working on urban development, when she was working as a municipal officer in São Paulo. When the Workers Party lost power in São Paulo in 2004, she was in Durban attending Brazilian–South African exchanges on urban poverty with South African SDI leaders, and – having seen it 'on the ground' – decided to set up *Internacao* to 'implement SDI ideologies here in Brazil'.

This snapshot from São Paulo reflects a central concern of this chapter: to show how learning is crucial to how SDI's urban connections function and develop. In this chapter, I argue that learning is central to how activists in urban social movements develop forms of organization and political

Learning the City: Knowledge and Translocal Assemblage, First Edition. Colin McFarlane.
© 2011 Colin McFarlane. Published 2011 by Blackwell Publishing Ltd.

strategy. I examine the politics of urban learning in the work of SDI, especially in relation to the Indian chapter, known as the Alliance, and argue, firstly, that central to SDI's work is a sociomaterial practice of learning in groups; secondly, that the nature of urban learning in SDI is key to the formation of political organization; and, thirdly, that this urban learning is critical to the construction of a particular political subjectivity of social change. I begin by briefly outlining existing work on learning and politics in social movement literature, highlighting a tendency in these debates to, on the one hand, stress the importance of information and knowledge in activism and mobilization but, on the other hand, to underplay the role of learning. I then describe how urban learning assemblages are formed and function in SDI through the example of tactical learning in model house construction. I go on to examine the learning of political organization in SDI through, in particular, its tactic of enumeration, or self-census, of informal settlements. The chapter uses the notion of assemblage as an orientation towards the agency of materials – including self-census maps, charts, enumeration documents and model house materials – in the learning practices of SDI's work, and shows the importance of documentary representations within urban learning assemblages as they travel through and give shape to multiple sites and practices. I end the chapter with a critical discussion of how SDI's learning practices around house construction and enumeration relate to and construct particular political subjectivities and conceptions of social change. If the previous chapter focused more on dwelling than on the translation and coordination elements of urban learning assemblages, this chapter will demonstrate the importance of translation and coordination.

Knowing Social Movements

Literature on global civil society tends to gloss over learning, with the result that we are left with the impression that knowledge travels straightforwardly between places in social movements and that learning occurs as a necessary consequence of the transfer of information or knowledge. This functionalist assumption of learning as knowledge transfer constitutes an important gap in the social movement and global civil society literature, as a few influential examples indicate. While Eyerman and Jamison's (1991) *Social Movements: A Cognitive Approach* sought to demonstrate the importance of knowledge in the views, interpretations, and social and political identities of actors within social movements, the book said little about learning practices. Kaldor's (2003: 95, 63) *Global Civil Society: An Answer to War* described 'transnational civic networks' as 'forms of communication and information exchange' and highlighted the 'importance of dialogue' between social movements in the East and West during the Cold War, but largely ignored the question of learning itself. Keck and Sikkink (1998: 24) highlighted the importance of communicative action in transnational advocacy networks – 'vehicles for communicative and political exchange, with the potential for mutual transformation of participants' – but offered few reflections on how knowledge travels and what the specific implications may be.

In his ethnography of anti-corporate movements, Jeffrey Juris (2008) went further in his examination of the rise of networking logics and practices internationally amongst activists. In one chapter, Juris argued that a crucial part of networking for activists is the communication of information. Building on Kevin Hetherington's (1998) idea of 'utopics', whereby radicals project their political ideals onto places, Juris (2008: 269) offered the concept of 'informational utopics' to 'characterize the way activists express their political imaginaries by experimenting with new digital technologies'. He showed how the Internet in particular is deployed as both imaginary and practice within 'horizontal' digital media activism. This rolecan vary from e-mail updates and rapid language translation ofcommentaries to websites and autonomous media projects such as Indymedia (www.indymedia.org). Established during the anti-World Trade Organization protests in Seattle in 1999, Indymedia was developed as an anti-corporate communications network involving translocal learning assemblages of electronic print, video, audio and photography, run and updated by activists, and for providing forums for exchanging ideas, information and resources. As Juris (2008: 270) argued, given that the central organizing logics of Indymedia have been direct democracy, egalitarianism and horizontal collaboration, it represents an example of informational utopics in which the logic of 'network' is both a form of technological organization and an activist imaginary.

Juris (2008: 286) argued that informational utopics do not represent a traditional utopianism 'involving totalizing visions of a far-off, perfectly harmonious world', but rather a 'postmodern utopianism' expressed directly, 'here and now, through concrete political, organizational, and technological practice'. The circulation of information clearly played a crucial role here for Juris, but social movement literature has not gone beyond this invocation of information and knowledge to consider how learning might matter in the activities, goals and organization of social movements. Infrastructures of learning play important roles in the emergence of new ideas and imaginaries in social movements, and if contemporary social movements constitute 'social laboratories' (Juris 2008: 297; Melucci 1989), then we need to be alert to the role that learning can play in generating new political practices and imaginaries.

As Leach and Coones (2007) argued, learning can be crucial to the shape of a social movement: the power relations between different groups in a social movement influences what knowledge takes precedence, what knowledge is marginalized, and how learning affects political positions over time. Two immediate forms of learning politics emerging from social movements are worth highlighting: firstly, the politics at work as different groups seek to define and delimit the object of learning. For instance, technical knowledge can become politicized and vice versa, conflicts over knowledge enable or hinder movements, legal knowledge may become a platform for resistance (Chapter 2), and the media can become a realm through which knowledge is discursively framed, translated, contested or simply ignored. For example, governments may define an urban water access debate as a technical problem and privatization as a technical solution, while social movements may tackle this very characterization of rendering technical (e.g. see Li 2008; Ferguson 1994, 2006).

Secondly, learning politics features through social movement campaigns around whose knowledge is deemed valuable and relevant. A lot of the activity of contemporary social movements involves lobbying for certain forms of knowledge to become central to the learning process. Consider, for example, the long history of Brazilian social movements campaigning for the participation of the marginalized in producing urban planning (see Chapter 4). In some cases, movements draw upon 'lay knowledge' and forms of 'experiential expertise' (Collins and Evans 2002) that people have acquired in everyday life. Increasingly, as Juris' (2008) work vividly reveals, the politics of knowledge are translated through cyberspace, for instance as movement participants gain direct access to governmental or scientific papers or volatile video footage and photography posted on movement websites or sent out to e-mail lists. Websites can serve as influential and far-reaching coordination tools for different domains of knowledge. But if clashes around knowledge are central to social mobilization, the battle-lines are often complexly drawn. Rather than government versus community or 'lay' knowledge, we often find 'discourse coalitions' (Hajer 1995) that cut across different groups that promote particular interpretations of problems and forms of learning. Urban NGOs, for example, may have allies in government, and internal fractures can position members on different sides of particular debates, as we shall see in relation to SDI's Indian Alliance. Leach and Coones (2007: 26) wrote of 'knowledge alliances' linked to different interests and power-laden processes, and argued for the need to attend to 'knowledge politics' in social movements, which 'interplay with the politics of struggle around material resources, and socio-political claims'. In short, they argued, 'contentious politics today is more often than not the politics of knowledge' (Leach and Coones 2007: 27). The challenge here is to unpack the different roles that learning plays within the work of social movements. I attempt this in relation to SDI in this chapter, revealing a complex and often crucial geography of learning in the movement.

Two words of caution here. Firstly, I am not suggesting that the response to the relative neglect of learning in accounts of the politics of urban movements should be to shift in the opposite direction to claim that politics *is* learning. The politics of urban social movements quite often does not involve any kind of learning. For example, the simple standpoint of 'no' – for instance, in 'stop the war' protests or gatherings against the demolition of urban informal settlements – does not itself embody learning (although the style in which it is communicated may do). Secondly, and following this, given the centrality of learning as translation–coordination–dwelling to urban movements, it quickly becomes very difficult to isolate learning in relation to the politics of social movements more generally. The relations between learning and politics must, then, be examined in the particular and contingent contexts of those urban movements.

Global Slumming

SDI is a collection of non-governmental (NGO) and community-based organizations (CBOs) working with urban poverty, particularly housing and sanitation, which operates throughout Asia, Africa and Latin America. It is a translocal experiment in building a new form of urban sociality; a learning movement based around a structure of what its leaders call 'horizontal exchanges' involving small groups of the urban poor travelling between neighbourhoods to learn from one another. The movement espouses a range of tactics that its leaders describe as indispensable to a development process driven by the urban poor. These include daily savings schemes, exhibitions of model house and toilet blocks, the enumeration of poor people's settlements, training programmes of exchanges, and a variety of other tactics. In seeking grassroots participation and horizontal exchange, SDI seeks to place urban learning at the centre of social and political relations. Its insistence upon learning from and between the urban poor emerges from the context of a failure – deliberate or otherwise – of the state to ensure collective provision of urban infrastructure, services and housing. SDI groups attempt to deal with the crisis of social reproduction in many cities of the so-called 'global South', from Mumbai (where much of the work of the network started) to Cape Town, Phnom Penh and Karachi. This crisis has been sharpened by the restructuring of the relationship between capital and the state in many cities in the global South, which has encouraged the privatization of infrastructure such as water or electricity supply, a general escalation of real estate prices, a growing trend towards gentrification, the collapse of various national and local welfare provisions, and increased forms of political repression in the form of demolition of 'slum' settlements to make way for land redevelopment (Smith 2002; Lorrain 2005; Davis 2006; McFarlane 2006b, 2008b,c; Wacquant 2007, 2008).

SDI's 'box of tools', as one leader in Mumbai described them to me, emerge from a complex range of histories of urban struggle in different localities, particularly struggles around preventing housing demolition and campaigns for security of tenure. These tactics are translocal in that they are place-based and emerge from particular local histories, but take shape through exterior interactions rather than being place delimited. They are not new in themselves, although they are inventive ways of thinking and doing and configuring and reconfiguring relations between actors (Chapter 1; Barry 2001); that is, invention names the process of assembling and reassembling that SDI's learning activities produce as the movement's tactics travel, and is a product of the interactions that constitute the assemblage. In this relational translocalism, the object of struggle remains the locality (for example, the local municipal corporation), and this is informed in part by 'the knowledge and tacit innovation of multiple other localities around the world engaged in similar localized struggles with similar local actors' (Sassen 2003: 11). As one SDI leader in South Africa, Joel Bolnick, has written (2007: 320): 'The [SDI] strategies (shared across the network) build on existing defensive efforts by grassroots organizations to secure tenure, and add to these existing efforts by measures designed to strengthen local organizational capacity and improve relations between the urban poor and government agencies.' SDI leaders argue that urban development strategies that are designed for the poor, but not *with* the poor, tend to fail: 'SDI believes that the monopoly over information and knowledge exercised by officials, technocrats and professionals needs to be broken and poor people themselves need to gain control over knowledge in order to deal more effectively with their situation' (Patel *et al.* 2001: 51–2). Its leaders articulate an entrepreneurial form of collectivist politics which emphasizes the capacities and skills of the urban poor.

This collectivist politics differs from Cumbers *et al.*'s (2008; Routledge and Cumbers 2009) discussion of 'Global Justice Networks' as bound by opposition to neoliberalism (and see Featherstone 2008). SDI's politics is less oppositional and is situated within existing local political economic frameworks through which it seeks to leverage space for the poor in urban planning and poverty reduction, a theme I will revisit towards the end of the chapter. In addition, we need to be mindful of the occasional romanticizing of translocal exchanges by SDI leaders and SDI publications that imply that cultural barriers have been seamlessly crossed. There have also been, despite the remarkable translocal reach of the movement, sometimes heated debates within the movement, for example when the South African Homeless People's Federation separated from SDI due to disagreements on political direction (Ley 2010).

The Indian Alliance – the Mumbai chapter of the SDI movement – is the most influential voice in the SDI movement and therefore a key influence on the form and politics of the urban learning activities that take place. It is a tripartite alliance of three groups based in central Mumbai: the National Slum Dwellers Federation (NSDF), a male-dominated activist movement that has a long history protesting and fighting demolitions and campaigning for basic rights; Mahila Milan ('Women Together'), a predominantly but not exclusively women's organization structured around local savings and loans schemes for emergency needs or collective housing and infrastructure investments; and the Society for Promotion of Area Resource Centres (SPARC), a NGO set up by middle-class activists in the early 1980s. NSDF and Mahila Milan share a resource centre that acts as a *de facto* headquarters for the national and international movement. The centre, in Byculla, central Mumbai, is a hub of activity: the three phones ring frequently and people from the local area constantly come in and out, some depositing money, some asking for loans, and some for advice from the NSDF individuals available. During telephone calls, as Arjun Appadurai (2002: 30) has commented based on his work with these groups, leaders 'exchange information about breaking crises, plans and news across these various locations in Mumbai – and also across India and the world ... a call [is] as likely to come from Phnom Penh or Cape Town as from Mankhurd or Byculla [in Mumbai]'. In its daily life, the centre is a space not just of support and solidarity, but of learning translocal political strategy and ways of working together.

Despite their commitment to learning through difference, some Alliance leaders position the Alliance as a dominant and more learned group within the movement, as one Mumbai leader said:

> What you have to do is see Mumbai as a hub that's like the crucible. All the new ideas [e.g. housing exhibitions, enumerations, savings] ... it's the most difficult place to work ... the size of the city, the scale of the problem. If you can solve something in Mumbai you can solve it in other places, and that's one of the reasons that we are not anywhere else.

On another occasion, the same leader referred to Mumbai as 'the mother base', while other Mumbai leaders have referred to the Alliance as a 'model' that is being adopted across SDI. This representation can be described as a form of power that, in John Allen's terms, is both manipulative and associational (see Allen 2003, 2004, on power; Chapter 5; on SDI and power, see McFarlane 2009a; and on power, space and social movements, see, for example, Featherstone 2008; Leitner *et al.* 2008). It is manipulative in that it is presented as a neutral set of facts and constitutes a simple message with extensive spatial reach. It is associational in that it involves an attempt to constitute a common agreement or shared will, i.e. that the Mumbai Alliance should lead the movement. While the Mumbai Alliance has certainly been the source of many of the strategies circulating in SDI, to say that it has driven or caused SDI activities risks creating an implicit hierarchy within the movement that sits uncomfortably with the state commitment to 'horizontal' exchange.

Through travelling encounters between cities as different as Cape Town, Phnom Penh and Mumbai, SDI's work is a relational product that combines, struggles over and negotiates the social and the material, and the 'here' and 'there', and the codified and tacit. The practices involved in constructing, adapting and putting its tactics to use is a process of learning through practice that sits alongside practices of lobbying, fundraising, state and donor negotiations, modes of solidarity, debates over political direction and whose knowledge counts, and so on. In exchanges, disparate forms of knowledges and identification circulate, from construction techniques to particular notions of the poor and social change. In short, 'horizontal exchanges' are translocal urban learning assemblages of materials, practices, designs, knowledge, personal stories, local histories and preferences, sometimes volatile debate, and an infrastructure of resources, fundraising, and state and donor connections. Here, assemblage places an emphasis on the agency of urban learning, on the bringing-together or forging of alignments between the social and material, and between different sites. SDI is a series of overlapping translocal assemblages that conjoin in different ways at different times around urban learning activities. For example, the South African Alliance, another SDI group member, has been very closely linked to the Indian Alliance over the past 20 years or so, and the relations between these two translocal assemblages has changed during that time depending on what was deemed important, whether in constructing model houses, or developing community toilet block designs, or discussing fundraising strategies or negotiating strategies with the state, or in planning how best to conduct local savings schemes within informal settlements. The next section will demonstrate the importance of learning to SDI's work through discussion of a key form of tactical learning in the movement that emerges from, but which also exceeds, incremental learning: housing construction.

The Housing Assemblage: Materializing Learning

In SDI, learning is first and foremost a sociomaterial practice of working in groups. A key example of this, and one central organizing strategy of SDI's, is the construction and exhibition of full-size model houses. This process of construction and exhibition, which incorporates designs by organizations operating within informal settlements, has circulated many of the cities in which SDI is based. Models and exhibitions hijack a middle-class activity (Appadurai 2002) and visibly dramatize the crisis of urban social reproduction, and are accompanied by group discussions ranging from concerns over land tenure to construction or local organizing. Models draw on domestic geographical imaginations and incremental knowledges of urban dwelling, and reflect a particular construction of the poor and of social change in SDI (McFarlane 2004, 2008b). In particular, they put the capacities and skills of the poor on public demonstration, creating an urban spectacle through which the poor are cast as entrepreneurial and capable

Figure 6 Model house, Cape Town, photograph courtesy of Astrid Ley

of managing their own development (a theme I will revisit later in the chapter). Figure 6 shows an image of a SDI model house held by a member of the South African Homeless People's Federation in Ekuphumleni, Cape Town.

Stories about how to construct model houses circulate SDI through an organized system of exchanges in which groups of the urban poor from different cities share ideas and experiences. In exchanges, visiting groups often join in with constructions and exhibitions as they are going on. Small-scale housing models, writes the Asian Coalition for Housing Rights (ACHR), an SDI partner, are often deployed as 'a three-dimensional imagining tool for people unfamiliar with the abstraction of scale drawings' (Asian Coalition for Housing Rights 2001: 13). The models – functional systems that enable learning through the coordination of different domains, including the home, aspiration, and a set of material pragmatics set in a context of urban poverty – are expressions of geographical imaginaries of the home.

ACHR go on to describe one exhibition in Thailand: 'As the model went up, the people pulled out boards, nailed things up differently, changed this, argued about that. Measurements altered, ceiling heights were raised then lowered, window positions shifted, bathrooms and kitchens swelled then shrunk' (ibid.). Models become the basis for negotiations around the kind of houses people want to live in, a process in which the collective will must be weighed against individual preferences, and which is subject to a range of social and cultural specificities and alterations. In model house construction, learning occurs through a relational process of body–group– materials, connoting less a form of contemplative dialogue and more a form of craft skill. This echoes Richard Sennett's (2008) argument that making *is* thinking – that transformation in both skill and self inhere in the practice of craft. The process of house construction in SDI is one that emerges through the experience of incremental urban learning as a specific mode of dwelling, but which through construction and exchange and exhibition exceeds everyday incrementalism and becomes a form of tactical learning, which intervenes in and opens up new possibilities for urban dwelling. In the process, model houses act as coordination devices as they are translated through exchanges.

In model house construction, learning is a process of achieving technocratic knowledge. For example, Mahila Milan women in Pune, near Mumbai, have trained themselves, with the help of a NGO called the Maharashtra Social Housing and Action League (MASHAL), in construction techniques pertaining to the building of toilet blocks. They buy materials, employ contractors, deal with corruption, engage in construction, and maintain the final product. Reflecting on how technical knowledge is learnt through translation, Amita Mbaye, part of a Senegalese Savings and Loan Network, said:

> When I asked the technician (who works with us in Dakar) to show us how layout plans are designed, he used such sophisticated jargon that I barely understood a word he said. In Protea South (Gauteng, South Africa) during our last evening, we asked a woman to draw us a plan. When she explained house modelling, I understood and felt that I too could do it. (Patel and Mitlin 2002: 132)

Model houses draw on the experiential knowledge of the poor for the purposes of housing construction. For instance, rather than focusing on conventional measurements, they stress non-technical knowledge. At the level of the individual house, construction techniques draw on the geographical imaginations of the poor (McFarlane 2004). Jorge Anzorena

(1987: 6), of the southeast Asian donor Selavip, highlighted the way in which the mechanics of housing construction become 'de-mystified' by drawing upon measurement techniques used by people who may be confused by conventional systems:

> [F]or example if 10 × 15 feet is taken as a standard size, then an average of 5 yard sari is 15 feet, hence one length would be the breadth of the house. A Mangalsutra [a necklace worn by married women] of 20 inches could be used for measuring 1.5 foot for a window. North Indian women could use the cord, which ties their Salwaar to get a measurement of 4.5 feet. ... Using various items which were on the women's person, such as a blouse length or a petticoat, all the structures within a home could be measured.

People are using their own knowledge created through experiences of constructing huts, using rudimentary materials including the clothes that they possess – rather than, for instance, tape measures that they are unfamiliar with – to begin the process of construction. In this example, learning occurs through the creation of a functional system – the sari or mangalsutra – that coordinates different domains, in this case, materials and the dimensions of the home, aspiration, and the realities and limits of urban poverty. The affordance of everyday materials is translated within assemblages that designate new urban possibilities.

Women play a particularly important role in relation to house construction, claimed Sheela Patel – a Mumbai-based SDI leader – writing with Diana Mitlin (Patel and Mitlin 2002: 158), because women have to bear the brunt of creating improvised 'systems to deal with water, sanitation and with delaying the frequent demolition of houses'. One specific example is women sharing stories about urban design and construction, a process which, for some at least, subverts the domestic burden women bear by using domestic knowledge in construction, as Patel and Mitlin (2001: 16) argued:

> Women can 'dream' about their houses better than men, in part because they use the structure that they live in more than men; men therefore easily concede that women can design the new structure better. Once that is achieved, men often concede that women who are trained to manage construction are the most effective supervisors of the process.

If this urban 'dreaming' is always learning-through-dwelling – in this case through the home as dwelling – women also 'teach' each other about what they have learned.

To be sure, this is a romanticized image that risks not only underplaying the complexities of gender inequality, but reproducing them by positioning women in relation to the domestic space and as a reference point for domestic labour. However, Patel and Mitlin (2001: 16) insisted that this is part of the overall process of increasing women's confidence and competence (capabilities, skills), thereby helping 'women to renegotiate their relationship with the community, with their family and with their husbands'. For example, writing about the experience of women in sanitation construction programmes with the state in Pune, Indian Alliance leaders (no date: 8–9) have written:

> On the job, they learnt new technical and financial skills. The city engineers and corporators took a while to believe that these 'Annandi' (illiterate) women as they refer to themselves were able to manage the construction, the finance, liaising with the city officials right from the level of the municipal commissioner, city engineers, and architects. They learnt to juggle with the not so cooperative corporators and in the process became more confident of handling the difficult situations they were put into.

Ideas from past model house constructions often inform future constructions elsewhere. Writing in reference to Mumbai's Railway Slum Dwellers Federation (RSDF) – an NSDF affiliate – the SPARC leadership (1988: 27) have stated that:

> [I]t was difficult to explain the dimensions of a 10 × 15 feet (150 sq. ft.) house to a group of people who have been living in 45 sq. ft. for most of their lives. ... However, life size models of houses designed by the people can be a very effective means of depicting the spatial dimensions. ... The RSDF committee felt that such an exhibition would facilitate their outreach among the rest of the railway settlements. There were four house models which were presented in the exhibitions, three of which had been built in the past by women pavement dwellers.

This story reveals how the house model operates as a coordinating device that emerges from the experience of urban dwelling but which translates stark and immediate concerns – the crisis of low-income housing in this case – into a form of tactical learning, a field of pragmatic opportunity that disrupts the everyday by disclosing possibility. As Patel and Mitlin (2002: 129) indicated: 'Through sharing their common experiences, community members identify solutions and make plans to instigate change. Gradually they start to identify new solutions to existing problems, passing on confidence and skills in the process.' House modelling in SDI is an assemblage of learning-by-doing that is at once social, practical, material, imaginative and translocal, and dependent on an ongoing series of translations between information, tacit practice, ways of doing and seeing, narratives of change, and circulating tactics.

If SDI's objective has been to gather attention, housing modelling has been a successful strategy, both at local sites where it has been hugely successful in attracting new members, media interest and policy-makers, and translocally where it has attracted considerable attention and funding from donors and international organizations (especially the World Bank and the United Nations, as well as the Clinton Foundation and the Gates Foundation). SDI leaders use geographical imaginaries of particular sites – for example, the home, in the shape of community designed models – as a basis for lobbying international organizations. One bold example of this combination was the construction by SDI activists of a full-sized model house in the lobby of the United Nations' New York headquarters during the 2001 Habitat conference. Arjun Appadurai (2002: 38) commented on the event:

> [Kofi] Annan was surrounded by poor women from India and South Africa who sang and danced as he walked through the model house and toilet that had been placed at the heart of his own bureaucratic empire. It was a magical moment, full of possibilities for the Alliance and for the Secretary General as they engaged jointly with the politics of global poverty. Housing exhibitions and toilets, too, can be built, moved, refabricated and deployed anywhere, thus sending the message that no space is too grand – or too humble – for the spatial imagination of the poor.

This translocal moment, however, should not be overstated or romanticized: SDI members were generally excluded or marginalized at the conference and the exhibitions are not representative of a shift in international politics. Sharma (2001: 9) argued that:

> ... the space given for the model house was not matched by a space for those who built it to be heard in the formal discussions. These were held in the basement of the UN building on issues that are central to the concerns of women like Rehmat and Shahnaz [pavement dwellers in Mumbai], and even the [homeless] women on the New York street. Yet, the structure of the meetings could not make space for the voices of these women to be heard.

That said, the importance of the moment for the movement itself should not be too easily brushed aside. In the exhibition, people from the margins of the contemporary urban landscape had come together to seize space at one of the centres of political globalization, a capture suggestive of what Keck and Sikkink (1998: 22) might have referred to as 'symbolic politics'. Having established the importance of learning through materials and in groups in SDI's work, the next section will show how particular learning tactics, especially that of 'enumeration', produce crucial forms of political organization in SDI.

Learning and Representation: Counting the Poor

Representations – specifically models, metaphors and especially documents – perform important roles in the learning of translocal political organization in SDI. Within SDI groups, formalized models and explicit metaphors are often put to use based on learning from experiences in other cities in which SDI operates. For example, in Bangalore, Thomas, leader of the city's NSDF group, told me that the organizational form of federation activities in the city followed a model codified in Mumbai with the Alliance that subsequently travelled through exchange. He pointed to four committees for managing construction projects as a specific example of this: (i) a labour committee; (ii) a purchasing committee; (iii) a finance committee; and (iv) a supervising committee. In Mumbai, years of Alliance experience led to the externalization of this model as a coordinating device that can be easily translated, and the model subsequently travelled as a particular institutional geography.

To take a different example from the Bangalore group, one member, Jula, said that exchanges with a group in Kanpur, Uttar Pradesh, had taught her how to 'do' savings though a simple, mobile model. The kind of competences she was referring to include the daily practice of savings, such as delivering individual passbooks to members, arranging groups of around 50 people into collection areas, and drawing up and compiling manual records. Lacksmi Shanmughan, of Bangalore District Mahila Milan (BDMM), referred to how, through exchanges with Mumbai, she and others learnt 'how they [in Mumbai] started their organisation through savings'. One practical example she gave was the use of improvised colour-coded money deposit boxes – for example, green for Rs 1 or red for Rs 2 – that helps organize the scheme and make it accessible. In these daily savings chains of translation, learning occurs through the experience of one group (in Mumbai) informing the institutional set-up of another (in Bangalore) in the production of a shared repertoire. The colour coding acts as a functional system that coordinates the domains of money collection, organizational ease, and the desire to familiarize the process for women who are often illiterate. These simple, everyday infrastructural competences are in turn altered through experience. For example, groups may draw on the representational form of daily savings but find that in practice it is more fitting in their own neighbourhood to have weekly or monthly savings than daily savings, perhaps due to earning patterns or a distinct culture of savings, as experience has shown in Hyderabad and in South Africa and Thailand, where saving is monthly rather than daily. On a different register, in another example of learning through coordinating devices, Asian Coalition for Housing Rights (2000: 29) listed ten 'tips' for exchanges, from getting a balance between NGO and CBO involvement, to presenting knowledge and ideas rather than attempting to convert people, to holding small, focused meetings rather than larger crowded meetings.

Metaphor is frequently used in relation to SDI's tactics, particularly in reference to daily savings, described in the Philippines as 'the building block', the 'common denominator' and 'the "glue" that holds communities together' (VMSDFI 2001: 74), and described by NSDF's Jockin in India as 'breathing', the foundation of the movement (Appadurai 2002: 34). Other metaphors used include cooking metaphors to describe the relations between knowledge and place. Alliance leader Sheela Patel has written (Asian Coalition for Housing Rights 2000: 8): 'The federation is kept alive by all this experimentation in all these scattered communities. It's like a hundred cooking pots simmering away, each with its own masala, its own concoction of local circumstances, personalities and whimsy.' This metaphor is both a way of translating the activities of the different SDI groups through codification, and of presenting an image of unity in diversity. It does so by anthropomorphizing learning – cooking, personality, whimsy – as the skilled experimentation of different ingredients (see Sennett 2008 on craft learning). As a potent image of SDI as a series of overlapping translocal assemblages, the cooking metaphor is an effort by SDI leaders to restate the unity of the movement through the vivid and universal compositional practice of cooking, and of encouraging SDI groups to view their circumstances as unique. This translation allows the SDI leadership to insist that different SDI groups do not need to be at the same 'level' and that they can be 'experimental', while underlining the central role of SDI's particular urban tactics for each member group – in other words, it is in image of translocal assemblages as co-functioning multiplicity rather than singularity.

The relationship between learning and representation is important not just because it influences the nature of knowledge creation, but because it is central to learning translocal political organization in SDI. One key example that I want to concentrate on here is the process of documenting enumerations. Enumeration is a tactic of self-census conducted by the urban poor on the urban poor. It was first carried out by the Indian Alliance in the mid-1980s as a response to the refusal of the state to enumerate pavement shacks. This lack of data informed erroneous state claims that the pavement population in central Mumbai was transitory or small in number, providing – if needed – another 'justification' (in the form of strategic ignorance) for state demolition of shacks. The purpose of enumerations in SDI is to organize its members and to create a space of negotiation with the state through the language of state policy, i.e. quantitative, mapped data of urban populations. As the Global Land Tool Network (GLTN) have suggested: 'Enumeration is a means to federate and organise communities and involve them in large scale slum upgrading projects' (2006: 4, cited in Huchzermeyer 2009: 4). Enumerations can also serve to pressure landlords into improving rental housing stock by exposing poor housing conditions. While enumerations can increase tensions between landlords and tenants, they can also serve to improve relations by leading to compromises between the two, as Huchzermeyer (2009: 17) wrote of one enumeration that SDI was involved with in Kenya: 'Whilst tenants now had the right to ask for services, landlords felt entitled to demand rent on time. As long as services were not provided, it had been hard to press for the rent.'

There is often a potential state audience for these data. For those in the state interested in improving informal settlements, basic data about living conditions are crucial. There is a growing awareness amongst international development agencies, notably UN Habitat, and amongst progressive state officials, that enumerations conducted by the state, rather than by communities themselves, often fail due to an inability or unwillingness of state officials to understand the diversity of neighbourhoods, or a lack of trust in state officials by the communities (Johnson 2007; Huchzermeyer 2009). Community self-enumeration, often referred to as 'peer evaluation' and usually combined with focus group interviews, is becoming more common as a result. Writing about an enumeration led by SDI-affiliate Pamoja Trust and commissioned by UN Habitat in Kisumu, Kenya, Huchzermeyer (2009) described the process of mobilization through organization that emerged. The enumeration, which included a wide array of groups including the Global Land Tool Network (GLTN) – an international network promoting 'grassroots' participation in land management – two community human-rights organizations, the municipal council (especially the GIS group in the Planning Department) and various other state organizations, had a range of impacts: it enabled community participation in state planning in some neighbourhoods, facilitated the development of issue-based mobilization around youth groups and the vulnerability of widows, and led to a greater awareness of the collective and diverse challenges people saw for the area. As part of this enumeration, SDI arranged an exchange group from Zimbabwe that consisted of not just Zimbabwean federation members and staff from the SDI-affiliated NGO, Dialogue on Shelter, but two Zimbabwean government officials – all of whom 'joined the team in order to transfer learning about grassroots enumeration to Epworth in Zimbabwe' (ibid. 2).

Enumerations are, of course, not without their drawbacks. For instance, enumeration can create data that are used by municipalities or landlords to improve revenue collection, clearly not necessarily in the interests of people living in informal settlements, unless the revenue collection is affordable and leads to improved services for the settlements. In addition, lack of coordination or capacities might mean that different neighbourhoods are enumerated more successfully than others, creating doubts about the data or piecemeal subsequent state interventions. Or, the data might simply be entirely ignored by the state or donor organizations. None the less, enumeration can play an important role as a learning tool that helps formulate translocal political organization. How might we conceptualize the construction of enumeration documents in learning translocal political organization? Documents – as 'paradigmatic artifacts of modern knowledge practices' (Riles 2006: 2) – in both their electronic and physical forms, play a crucial role in organizing, framing, narrating and contesting urban life. While they are often reduced to mere 'supplementary data' or 'background' information in urban research, they are important examples of how urbanism is learnt and being learnt about.

Data-urbanism

A range of heterogeneous materials are engineered and mobilized in learning through enumerations. The tactic of enumeration vividly demonstrates that learning and the sites through which learning is produced are not separate, but mutually constitutive. The concept of translation discussed in Chapter 1 highlights the intermediaries, sites and practices through which urban learning is heterogeneously produced, and calls for attention to the material geographies involved in its production. For Bruno Latour (1999), an important consequence of this is that 'things' become engaged in discourses through a series of translations. These discourses can ultimately come to 'take the place of the original situation', just as – in one of Latour's accounts – a diagram of a forest can come to take the place of a forest itself (Latour 1999: 67), or a map of or set of demographics about an enumerated settlement may take the place of the settlement itself.

This 'material semiotics' (Akrich and Latour 1992) is 'concerned with how all sorts of bits and pieces – bodies, machines, buildings, texts – are associated together in attempts to build order' (Bingham 1996: 643). We might think, for example, of Annelise Riles' (2001) account of the knowledge practices of Pacific Islanders in UN conferences, which attends to the agentic role of documents, funding proposals, newsletters, and organizational charts, in the institutional life of organizations, and which echoes Holloway's (2000) and Philo and Parr's (2000) attempts to conceptualize institutional geographies in travel, drawing upon, amongst other traditions, actor–network theory. As they argue, institutions are never 'pre-given' or 'formulated in advance', but are constantly in the making, and act in multiple ways through the weaving together of heterogeneous materials – as 'active space-times in allowing certain knowledges and practices concerning ourselves and the world to be performed and made to travel ... [they are] always becoming stabilised, rather than already stabilised and formalised relational effects' (Holloway 2000: 654). This focus on the distanciated and sociomaterial making of SDI through, for example, travelling documents, is a useful way of thinking about the role that learning to make documents plays in political organization.

An example here is the production of the Indian Alliance's document, *We, the Invisible* (*WTI*), produced in the mid-1980s. *WTI* represents the first enumeration conducted by the Alliance, and by implication the first conducted by SDI. It is a collection of data based on an enumeration of people living on pavements in central Mumbai in 1985. Pavement dwellers are frequently and variously constructed as spaces of 'encroachers', 'illegal', 'transitory', 'dangerous', 'subversive' and 'immoral'. They tend to be overlooked by state planning and policy development, as Mohapatra (2003: 298) argued:

> The state has never included them [pavement dwellers] in any census 'as a matter of policy', because it views them merely as intruders into or encroachers of private/government property. Thus, the state's refusal to recognise them resulted in denying the pavement dwellers their status as citizens ... The only way the state approached them was by dislocating them, demolishing their hutments, demonising them and by keeping them away from its network of entitlements and rights.

It is often claimed by authorities that pavement dwellers are highly transitory and this claim can be used to justify demolition, or a lack of ration cards and other essential services and infrastructure. The Alliance conducted the survey along with local people who volunteered, and it brought many of the pavement dwellers involved into new understandings of their pavement settlements. Using a municipal corporation map to mark out local boundaries, people walked street by street in concentric circles, marking high-density areas of pavement slums into 'clusters'. They produced a cluster profile. In simultaneously bounding and interlinking different territories, representing places as clusters actively creates a translocal organizational field. Many people in these new territorially-bounded clusters were organized through local meetings into Mahila Milan groups. The clusters were systematized into charts and graphs and the process of building the document on computer began. Reflecting on the process, two Alliance leaders discussed the frustrating, inherently sociomaterial learning process of making documents by commenting on the role of computers:

> Since none of us had any training in computers, we were totally mystified by the 'computerese' spoken by our EDP [a computer firm] friends as they no doubt were by our 'activese'. We were told the size of our data – 6000 households and almost 27,000 individuals – was throwing up a host of problems for a programme normally designed to handle a much smaller volume of data. By the time the programme was 'debugged', there was no time for anything but simple frequency tables (Batliwala and Patel 1985: 13; and see Riles 2001).

In both *WTI* and a second enumeration report concerning hutments along railway tracks produced by the Alliance in the late 1980s – *Beyond the Beaten Track* – a series of translations take place to enable the codified distilling of knowledge, experience and histories into documents. These translations involve what Latour (1999: 67) called 'mobilisation': the enrolment of heterogeneous materials, including paper, PCs, cameras, films, developing materials, printers, maps, questionnaires, local people, resources, and so on. The enumeration included 6000 households in central Mumbai, and the results were documented in *WTI*. The document included graphs with information on, for instance, length of stay in Mumbai, pre-migration occupation, and state of origin. Contrary to the popular belief that pavement dwellers are transitory and mobile, over 60% had lived on the pavement for more than 12 years, and many were born and brought up on the pavements. Alliance activist Sheela Patel (no date: 203) has written:

> The results of the survey were dramatic: most pavement dwellers were from the poorest districts of the country, they were landless agricultural labourers and artisans who had no property or assets in the village, they had come over 20 years ago (in 1985), and more than half the population (more than the national average) worked, and yet earned less than a minimum wage. Most walked to work as a means of subsidising their transport costs, so staying near the place of work was essential.

The document sketched the lives, histories, jobs and movements of pavement dwellers, and the relationship between pavement dwellers and existing law, policy and institutions. The work of documentation represented an act of what James Holston (1999: 167; 2008) has called 'insurgent citizenship', in that it sought to introduce new citizenship identities 'and practices that disturb established histories' of citizenship. The codified knowledge in *WTI* is of a form that *discovers* what is already there but usually excluded as much as it *constructs*, for example, through arguments and diagrams, *invents* new possibilities for slum improvement, *subverts* the census as a tool for interrogating mainstream imaginative geographies of pavement dwellers, and *insurges* in attempting to create new spaces of engagement with the state. These different modes of knowledge stem from a series of sociomaterial translations that create statistical, quantitative knowledge by poor people that speaks the language of the state. This process of creating codified knowledge can be a process of political organization, as Johnson (2007: 288) wrote of an enumeration in Uganda: 'The survey was at the centre of discussions about actions to be taken by households and the local council to improve water, hygiene and housing conditions. It temporarily promoted further codification of knowledge in terms of generating records of systematically gathered data.' Enumeration documents are produced by learning through chains of translation and enable learning through the coordination of multiple domains, from maps and statistics to legal debates, state policies, and the possibilities for political action.

As an example of tactical learning that stands apart from and intervenes in the possibilities for urban dwelling, the *WTI* enumeration facilitated two other important changes for the Indian Alliance groups. Firstly, it organized pavement dwellers into Mahila Milan cooperatives. For Sheela Patel (no date: 3), this was the most significant impact of the enumeration: '[T]he communities ... as a result of this exercise, now began to see themselves as a group with common needs and aspirations and began to explore the possibilities of organising themselves.' The creation of clusters and Mahila Milan cooperatives was a re-imagining of local geographies, and in these re-imaginings, new possibilities for politics emerged. People discussed different possibilities for their livelihoods, including plans for housing and collaboration with government authorities. Each step in the chain of translation towards producing a document of the enumeration was a process of learning political organization for the Alliance.

Secondly, *WTI* was part of a broader campaign to form a negotiating platform with the authorities, representing, as Patel has put it, 'a tool through which they [pavement dwellers] were talking to the rest of the city' (no date: 3). The Mahila Milan groups – with the assistance of awareness-raising sessions run by the Alliance at Sunday meetings in the (recently acquired) Area Resource Centre on, for instance, ration cards and rights – attempted to pressure local authorities, as this statement from one prominent Mahila Milan member, Lakshmi from Byculla, central Mumbai, indicates:

> The municipality used to come and demolish [our huts] everyday. Now that is stopped since we formed our Mahila Milan. Earlier they did not make ration cards for us, they said that the person living on footpath does not have any right to a ration card. Now, after coming into Mahila Milan, we have become brave. Now we talk to the senior, to the inspector, everyone in the police station. We have learnt this after Mahila Milan came. (Mohapatra 2003: 305)

Each person received a copy of *WTI* in their own language, and the document was sent to local, state and national decision-makers. It also travelled more widely, gaining attention among donors and professionals, for example, or through SDI exchanges. Alliance members who have conducted enumerations explain the history of documents and the steps of enumeration as parts of learning about enumeration during exchanges. The pavement enumeration was later mobilized in the railway enumeration as a codified example of how to conduct enumeration. But whatever role these documents play in learning within different contexts, they do not have any pre-given agency but instead are put to work in specific contexts of interaction (by NGOs in SDI, by particular community groups, by donors, by state officials), constituting new chains of translation. In the creation and subsequent circulation of documents, the lines between representation, performance and context collapse in the practices of translation–coordination–dwelling in which they are located.

In the travelling of enumeration, pieces of paper become standardized and form a basis for simplifying and collecting data about settlements. Standardization in enumeration is most visible in the enumeration form itself, which is designed to compartmentalize data about the settlement as a self-demarcated cluster. For example, members of the Bangalore Slum Dwellers

Federation showed me a standard slum enumeration sheet they use, which compiles data on a settlement's age, boundaries, land ownership, authorization, electoral and taxation details, as well as information on the population, number of houses and families, and details on the caste constitution of the area.

As a tool for political organization in SDI's work, the 'slum' enumeration sheet is what Latour (1999) would call an 'immutable mobile', a coordinating tool between different forms of knowledge that can travel translocally with relative ease through a series of translations. Settlements are mapped, houses are enumerated, and are transformed into a mobile collection of data later analysed by community organizations, in neighbourhood houses, and even occasionally in the offices of the Municipal Corporations. When the maps and enumeration statistics are accompanied by written text – as in, for instance, *WTI* – the data reveal not just codified knowledge and trends not known previously, and which could not be known without the enumeration, but features of settlements often unknown to inhabitants themselves. In announcing particular features of the urban locale, responses and scenarios are subsequently imagined and discussed. In this sense, pieces of paper can encompass the sites – divided into clusters – that, paradoxically, disappear even as the observers sit and analyse them in the settlements. In this reversal of space-time, 'we are able to oversee and control a situation in which we are submerged, we become superior to that which is greater than us, and we are able to gather together synoptically all the actions that occurred over many days' (Latour 1999: 65). As Latour wrote of pedologists researching the Amazon forest, it is only when samples are abstracted for analysis from the dense confusion of the forest that rapid progress is made: 'In losing the forest, we win knowledge of it' (ibid. 38). Documenting enumerations is a simultaneous process of spatial discovery and creation, and one that creates a particular kind of control through the creation of a translocal organizational space which is actively assembled as *known* in certain quantitative ways.

Enumeration involves learning a particular political organization of slum settlements. It is a device of learning through coordination, in that it is a functional domain that centralizes knowledge about jobs, occupancy, ration cards, and migration, and in its operation as a centre of calculation for discussing possible urban futures. As it travels translocally, the enumeration document translates into new contexts amongst states or donors, or other SDI groups, and both intervenes in and opens up possibilities for urban dwelling. The enumeration tactic has moved and changed through India and cities as diverse as Cape Town, Victoria Falls and Phnom Penh. For instance, Indian Alliance leader Jockin Arputham started planning an enumeration in Mbare, an inner city area in Harare, Zimbabwe, with local federation leadership during an exchange: 'The Zimbabwean Federation would survey all 30,000 residents and in so doing mobilize them into the new federation' (Chitekwe and Mitlin 2001: 92). The Namibian Federation, who were visiting Zimbabwe, helped in this enumeration, learning the process as they went and sharing the tactic with others (ibid. 93–4):

> Forty Federation members from around the country would gather in a small office borrowed for the duration. Several groups took questionnaires around the blocks of flats. One group of some ten people remained in the office, collating the results by hand and aggregating the findings. Residents came out to find out what was happening; they were curious about the questionnaires. Many were excited when they found out that this was being done by residents. ... Nightly meetings introduced residents to the idea of savings schemes and explained how they too could organize themselves. Within two weeks, 12 savings schemes had been formed.

Enumerations can directly encourage mobilization around issues that emerge from it. For example, one respondent in the SDI enumeration in Kisumu, Kenya, told Huchzermeyer (2009) that savings schemes emerged from the enumeration that would lead to construction of local toilets. Enumeration documents and practices allow SDI groups like the Alliance to '(potentially) exert an influence, to be able to act across space and through time' (Holloway 2000: 557) in attempts to influence authorities, organize people, and mobilize new political possibilities. Documents travel through public and private organizations, donors, and other NGOs and CBOs as a result of – to borrow from Holloway (2000: 557) – three processes of transformation involved in chains of translation. Firstly, *simplification* – the complexity of information about the lives of those involved is 'suppressed'. Secondly, *discrimination* – the documents brought into being are 'now clearer objects' than what they replace. That is, they are a particular view of the slum, charted in frequency tables, graphs and pie charts. Thirdly, *interrelation* – although different in nature, the documents are understood as similar objects to other documents, and in particular the quantitative knowledge of government census data. Through documents, the Alliance's ideas are able to move at new speeds and distances through a range of existing and emerging translocal urban learning assemblages, although the knowledge is variously received and interpreted. As codified representation, the knowledge can move more rapidly than tacit knowledge. The documents are posted on the Alliance's and SDI's websites and reach greater distances and take on new mobilities, in the process changing form and becoming more durable while simultaneously more available and limited (people living on Mumbai's pavements, for example, are on the whole less likely to be able to go to the website).

Through these multiple space-times of urban learning, new knowledge about settlements emerges from analyses of data, which emerges from the compiling of maps and tables using various materials (statistical packages, computers, etc.), which emerges from data collection by whatever volunteers can be mobilized throughout the area, which emerges from questionnaires, which emerge from the idea of conducting an enumeration, which emerged from the exclusion of pavement and slum dwellers from government censuses. In different places in SDI at different times, other translations feature. For instance, an enumeration in Kisumu, Kenya, consisted of parallel focus groups that led to compiling a list of people's priorities, and in the same enumeration data were verified not through door-to-door confirmations, which SDI leaders promote, but through mobile phones facilitating quick and immediate verification of information, and through mass meetings with PowerPoint© presentations showing household data and physical mapping (Huchzermeyer 2009). Latour has referred to such multiple translations producing material representations as 'inscription': a general term that 'refers to all the types of transformations through which an entity becomes materialized into a sign, an archive, a document, a piece of paper, a trace' (Latour 1999: 306). These inscriptions – particular forms of coordination devices such as graphs in enumeration documents – are quantitative, empirical representations, and SDI has found they can gain credibility with local states as such. Enumeration subverts the more general notion of census as a tool of governmentality (Foucault 1978), as Appadurai (2002: 36) has noted. It is subversive in that while it administrates a population and creates data about it that can be managed by the state, enumeration in the hands of SDI is a means for attempting to organize, alter the terms of political engagement with the state, and negotiate for rights.

Statistics here are less an instrument in the government of populations, and more an instrument for enabling the poor to mobilize themselves as a population and influence authorities. This data-urbanism represents a means to try to create space for activists in urban planning, as one Alliance leader, Sheela Patel (no date: 2), has argued: 'Statistics are means through which planning is done, and cities, especially those in the south, tend to plan for the citizens who live in formal housing, attempting to ignore, and thereby not cater to, those who live in informal settlements.' And following the survey, she continues: 'Interestingly, in the past all possible resources allocated by the state to the poor excluded the pavement dwellers. Now the demands for their inclusion began, because the survey also demonstrated that they were indeed the poorest and the most vulnerable' (ibid. 3). This recognition extends beyond the state to 'researchers and practitioners at institutions and NGOs, who had hitherto not seen these people as a 'population' ... and [now] increasingly began to include the pavement dwellers as a category among the poor' (ibid.).

The politics of learning function here as a play of cantering and decentring knowledge. In Foucault's (1978) lecture on governmentality, he described how census statistics in the seventeenth century rendered possible the decentring of the notion of the family and the centring of the economy for states. He argued that statistics began to reveal that:

[P]opulation has its own regularities, its own rate of deaths and diseases, its cycles of scarcity, etc.; statistics show also that the domain of population involves a range of intrinsic, aggregate effects, phenomena that are irreducible to those of the family, such as epidemics, endemic levels of mortality, ascending spirals of labour and wealth; lastly it shows that, through its shifts, customs, activities, etc., population has specific economic effects: statistics, by making it possible to quantify these specific phenomena, also shows that this specificity is irreducible to the dimension of the family. (Foucault 1978: 99)

The creation of a statistical population though the Alliance's act of self-governmentality is a particular process of relearning the 'urban settlement' by simultaneously decentring the elitist idea of a scrambled mass of temporary households, and centring a range of mapped clusters, with data attached, that the Alliance can then attempt to mobilize in negotiations for housing, sanitation, ration cards, water and sanitation provision, garbage collection, and policy formulation. Documents like *WTI*, then, are 'quasi-objects', as Michel Serres might describe them – they can act in various ways and are therefore more than what the traditional sense of 'object' allows (Serres 1974; Serres and Latour 1995; Bingham 1996). The strategy of enumeration continues to be translated through SDI exchanges. For example, in Brazil, Anaclaudia, the NGO leader newly linked to SDI mentioned at the start of the chapter, told me that enumeration was the most exciting aspect of SDI's work she was hoping to use: 'Bringing information is so important ... to discussion ... otherwise you keep having empty discussions, you cannot diagnose it.' Enumeration, then, is important to how SDI groups produce and learn to organize their urban political geographies, and to how they attempt to negotiate with the state. But what sorts of political subjectivities are constructed in this urban learning? And what is at stake in them? The next section steps back from the detail of house models and enumerations to critically engage with how this kind of tactical learning relates to the construction of political subjectivity and social change for SDI.

Entrepreneurial Learning

As tactics for learning political organization, model house construction and enumeration play a key mobilizing role for SDI groups and often lead to savings groups for housing or sanitation. Central to this is the expression of a particular subjectivity of the capable, learning poor. As Huchzermeyer (2009: 14) reports from an SDI enumeration in Kisumu, Kenya:

> The importance of mobilisation for self-reliance, as promoted through the SDI approach, was explained by a Muungano member directly involved in organizing the slum enumeration. He noted the prior attitude of the

community of expecting per diems for any meeting attended. This had to change to an attitude of voluntary participation, which was a basic requirement for a grassroots enumeration exercise to be carried out.

SDI is not seeking to politicize urban development through street protests or through attempting to invoke constitutional rights (such as the right to property) in negotiation with authorities, but through demonstrating the *capacities* of the poor in learning practices. Exhibitions are practical illustrations of the capacities of the poor to build houses, while enumerations are illustrations of the ability of the poor to conduct a census through a particular kind of settlement learning involving walking streets, interviewing people, organizing clusters and groups, and compiling maps and graphs. These learning practices are central to the political identity of the poor and the conception of social change in SDI's work.

The Indian Alliance, for example, seeks to re-imagine the poor as a skilled, capable group of people – in short, as active *learned citizens*. This contrasts with the imaginative geographies of the urban poor as criminalized encroachers often popularized in Mumbai. It is a conceptualization of the poor connected to the Alliance's imaginary of social change, which asserts that this skilled group should be managing their own urban development, and in doing so resonates with the controversial entrepreneurial conception of informality offered by Hernando De Soto (2003) of creating mortgageable individual titles that release the 'dead capital' of informal housing. One alternative to the Alliance's – and SDI's – conceptualization of the poor would be the political construction of the poor as an exploited class, for instance as poor wage labourers, or as a set of classes (Patnaik 1999), with labour power it can use or withhold. Why is the Alliance – and indeed SDI more generally –mobilized around the first concept? Why, for example, have people not been mobilized around struggles for living wages, employment security, free education and free healthcare? Why is the mode of political subjectivity one of the patient learned citizen working with the state rather than, say, the radical protesting agitator?

Part of the answer here is simply that it has been the desire of the majority of Alliance members to negotiate with authorities rather than to protest by, for example, withholding labour, or taking to the streets over free healthcare (Das 2000). Learning is crucial to the cultivation of this political style and strategy. As one international supporter of SDI commented, when pavement dwellers were threatened with eviction, SPARC's response not to demonstrate and publicly fight was based on discussion between SPARC and the Mahila Milan groups. Mahila Milan's preference was to negotiate.

There is a particular conception of social change that is emerging from the Alliance, one that mobilizes the poor around their own skills and capacities, and that seeks to build collaboration with government officials through demonstrating these capacities. Much of this emphasis hinges on a discourse of non-party alignment. Non-party alignment refers not to a withdrawal from politics but to an undiscriminating engagement. It is a strategy of negotiating with whoever is in power, of avoiding party politics, avoiding ideological debates, and seeking out collaborations and partnerships with the state. In part, then, this is a strategy for attempting to avoid co-option and clientelism. This also marks a historical shift for many SDI members from oppositional politics and street protests against the state to negotiating roles with the state. One urban development specialist associated with SDI described this approach to the state as indicative of a shift from politics with a big 'P' to politics with a small 'p'. For the Alliance, it amounts to a 'politics of accommodation, negotiation and long term pressure', what Appadurai (2002: 29) referred to as the movement's realpolitik or 'politics of patience'.

One SDI affiliated individual reflected on the remarkable commitment to this process across the SDI movement:

> There was a row in Victoria Falls because one of the leaders used the federation office for Zanu PF organising during the Presidential election, all the other members got pissed off. So there are abuses by particular individual leaders, but they are the exceptions and even this particular guy did not go to the National Executive and say 'we should support Zanu PF'. ... Political parties will try and use the federation, certainly, and the federation may allow itself to be used but it's equally happy to be used by the next political party. ... I don't think any national federation leadership has argued that it should be party-aligned.

This approach, which Diana Mitlin (2008) has called 'co-production', has proven incredibly successful. The Indian Alliance is one of the most influential civil society collectives in India on urban debates, and the funding they received for their leading role in Mumbai's World Bank-sponsored Slum Sanitation Programme – to build, originally, 278 toilet blocks across 20 wards in the city – constituted the largest allocation of developmental work by the state to an NGO in Indian history (McFarlane 2008c).

SDI more generally has become influential amongst multilateral donors, especially the World Bank, and bilateral donors, and in 2007 received US$10 million from the Gates Foundation to support its development of Urban Poor Fund International. For example, in a Cities Alliance meeting in London in 2007, representatives from the Cities Alliance, the Ford Foundation, the Gates Foundation, the Rockefeller Foundation, the government of Norway, the United States Agency for International Development (USAID), and the UK Department for International Development (DFID), met to share detailed information on individual programmes of support to SDI and to discuss ways in which the organizations represented could better collaborate with each other and improve the quality of support to SDI.

The combination of an entrepreneurial conception of the skilled poor and the discourse of non-party alignment has been the source of some criticism for the Alliance in Mumbai. It has been criticized by other NGOs and political figures in Mumbai for a perceived politics of accommodation, with some suggestions that in practice the approach has led to a masking of the causes of urban poverty. For example, many NGOs were sceptical of the Alliance's support for free housing for informal settlements in the controversial 1995 Slum Rehabilitation Authority Scheme, on the basis that it would give the impression of solving the problems of urban poverty while leaving the structural conditions of the exclusionary political economy of real estate in Mumbai unchecked. This scheme funds free apartments for the poor through private investment, and involves civil society organizations in the moving of poor people. Free housing was a controversial election promise from the Hindu fundamentalist Bharatiya Janata Party (BJP)–Shiv Sena alliance. The scheme requires no financial participation from the government and is designed to pay for itself. The free apartments, built as (often substandard) tower blocks, are cross-subsidized through the sale of around half of the constructed apartments on the open market – an example of the increasingly common strategy of land-sharing in urban poverty debates. It is a capitalist entrepreneurial housing strategy, and has been severely criticized by a range of voices, including academics, NGOs and political figures. For instance, Banerjee-Guha (2002: 124) has written:

> An outright real-estate oriented proposal, this scheme intended to use land as a resource by offering additional FSI[1] in excess of 2.5 to attract private builders. ... The success of the scheme depended on high property rates, to benefit large developers and builders who thus became instrumental in evacuating slums.

The director of another Mumbai-based NGO working with urban poverty noted SPARC's involvement on the board of SRA, and suggested that '[SPARC are] not always representing the best interests of the slum dwellers'. These criticisms were echoed by one official at the Maharashtra Housing Board. He described the cross-subsidy scheme as 'ill conceived, badly formulated, wrongly drafted – a recipe for disaster'. He elaborated that the Alliance:

> ... owed responsibility to slum dwellers [and it supported a plan] that ... was entirely for the purpose of the builders ... but they said 'as long as there is some scope for NGOs to be incorporated in the scheme then let them [the

---

[1] Floor Space Index (FSI) refers to the relationship between location and floor space. The maximum floor space for a house is 225 sq. ft in a rehabilitation tenement (for the poor), but there is no restriction in a sales tenement (for sale on the market). FSI is 2.5 for *in situ* construction. If 2 FSI are used for the construction of rehabilitation tenements then 0.5 must be used privately to subsidize the development.

state and the builders] have their own agenda'. ... Perhaps she was right because she realised she couldn't do anything beyond that, but I thought that was a major set-back for the whole movement of slum dwellers.

He added that despite the protests of every other NGO associated with urban poverty in the city, the Alliance's support has meant the waste of a potentially critical voice that could have raised important issues and challenged the hegemony of the builders at a critical time. Instead the 'system', he asserted, remains in the long term. In his view, the Alliance should have campaigned for the regulation of builders – which, in his opinion, is what some other NGOs in the city working with slums do and would have done if they were in the Alliance's position. However, as one influential SDI supporter retorted:

> SPARC [the Alliance NGO] can hardly have let down the slum dwellers movement when NSDF and the slum federations that are within it have chosen to align with SPARC and work with them. The relocated families do face difficulties in their new sites but they would not return to live on the railway tracks if given the option and there is a pride in how they managed the process and in their new apartments.

Similar criticisms have been made of the South African Homeless People's Federation (SAHPF). As Huchzermeyer (1999) put it:

> The Homeless People's Federation professes to be a pragmatic rather than political movement. While its achievements in securing the uTshani Fund agreement [a government fund for housing subsidies] in various provinces must be highly acclaimed, the apolitical nature of the Federation does mean that it lobbies primarily for concessions rather than a fundamental review of the policy on housing and governance. (Huchzermeyer 1999: 72)

Since 1999, the SAHPF has become divorced from SDI and its South African NGO representative People's Dialogue, the South African equivalent to SPARC.

In this reading, both the Indian and South African Alliance's commitment to non-party-alignment amounts to working within rather than against mainstream political frameworks. The kinds of political engagements that are emerging from the Alliance's urban learning around its tactics of enumeration and house model construction are contested, and pose some difficult questions around the politics of its learning strategies. In the short term, the Alliance's entrepreneurial approach is certainly leading to some cases of improved housing and sanitation conditions. In the longer term, the Alliance is involved in the development of poor people's confidence and abilities, in challenging the politics of patronage that often constitute relations between the state and informal settlements, by creating and mobilizing organizations of the poor around negotiations with the state, and attempting to build a process of long-term transformative change in gender relations. But this approach is not striving to radically change the long-term politics of private and state interests, and may even be serving to entrench them.

The Alliance, then, is working with symptoms of poverty, in that it is not engaged in radical long-term structural changes in the control of builders and developers over resource distribution in Mumbai. The movement is not attempting to structurally alter the course of development in the city through housing schemes like the SRA, but instead is seeking to ensure that programmes such as the SRA scheme are implemented with a strong element of people's participation. However, the Alliance has made some substantial gains in Mumbai for the poor, and whatever its realpolitik loses in terms of keeping radical urban change on the agenda, it may well represent a more plausible general approach for poverty alleviation than the more oppositional approaches of some other NGOs in the city. The Alliance defends its position in reference to a discourse of collective empowerment and control. One Alliance leader situated the Alliance's emphasis on agency in relation to the relationship between structure and agency, arguing that a successful negotiation between the two can only result from an organized collectivity:

> All of this is a new way of thinking and doing ... obviously none of us have complete control over our lives. The larger circumstances in which we live dictate these things. But different circumstances make us feel more comfortable, more in control, more able to dialogue and negotiate ... options increase, choices expand. All these are ... variables that contribute to our sense for having more control versus less control. So what the federation philosophy is, is that poor people cannot expect that kind of control out of some individual empowerment, it has to come out of collectivity.

Conclusion

Urban tactics like model house construction and enumeration add, to borrow from Raymond Williams (cited in Harvey 1996: 30), an 'extraordinary practical vividness' to the prospects of individual and common improvement. They do so by, firstly, focusing centrally on a sociomaterial practice of learning in groups as a basis for reimagining housing options; secondly, creating, through enumerations, an urban geography within settlements that is also a process of learning political organization, especially through enumerations; and, thirdly, in combination with a certain realpolitik of non-party alignment, contributing to the cultivation of an entrepreneurial conception of the learned citizen negotiating in partnership with the state. Tactics like enumeration or model house construction are examples of tactical learning through coordination devices which are produced and travel through multiple domains of translation. As forms of tactical learning, they enable participants to learn forms of political organization, new possibilities for urban dwelling, and ways of negotiating with the state. These learning activities cannot be separated from the politics of SDI. SDI leaders negotiate the possibilities of urban political change through the construction of active, learning, skilled citizens who are portrayed as only needing the land and opportunity to manage their own urban development. If political organization is learnt, political subjectivities emerge centrally from those learning practices.

SDI's urban learning assemblages dramatize the crisis of urban dwelling, whilst attracting in equal measure significant positive attention and funding, and forceful critique. For some of the Indian Alliance's critics, and indeed critics of other SDI groups, the movement's realpolitik is too ready to make sacrifices for state dialogue. Rather than a political subjectivity of the skilled citizen seeking partnership with the state, SDI's critics would seek alternative subjectivities: for instance, the dispossessed informal movement not seeking to learn with the state but to demand basic rights from it, or, instead, refusing to become formally involved with the state at all for fear of political cooption in the name of learning partnerships. This is a theme I will expand upon in the next chapter when considering the nature and politics of learning within and between state and civil society through the notion of 'urban learning forums'. For both the movement and its critics, at stake in these increasingly influential learning politics are the future prospects of radical informal urbanism.

In making learning an explicit and central part of its activities, SDI acknowledges what many accounts of social movements fail to account for: the central role of learning in the activities, organization and political strategies of social movements. Rather than functioning as a supplement or appendage to social movements, learning is crucially woven through their activities and politics. SDI is not unique in this sense, even if it places more emphasis on learning activities than other social movements. All movements must learn: learn how to work together, how to lobby the state, how to develop objects of struggle, how to process often overwhelming quantities of information, how to use resources, how to avoid strategic mistakes, how to conceive the nature of urban politics and to intervene in it, and so on. One of the key contentious areas here is the politics of learning between civil society and state, and it is to this that the next chapter turns.

Chapter Four

Urban Learning Forums

Introduction

In this chapter I examine the sorts of environments that might give rise to progressive forms of urban learning by developing the idea of *urban learning forums*. The forum is a particular type of urban learning assemblage in that it signals the production of a centralized and organized environment specifically geared towards learning between different actors, including the state, donors, non-governmental organizations, local groups, researchers and activists. It is a specific space-time within learning assemblage: an organized encounter that may be a one-off or part of a series of events. The forum is an example of learning through coordination in that it centralizes and translates multiple different forms of knowledge from different people and groups. I pick up on some high-profile examples of learning in participatory urban forums such as those in Brazil, and examine the role of different actors in these processes. The chapter argues that the success of participatory urban learning forums depends upon, firstly, and drawing on Callon *et al.*'s (2009) study of science and technology controversies, the *intensity, openness* and *quality* of these forums; secondly, the commitment of state authorities to the participation of the marginalized and the poor and to ceding decision-making powers to the forum; and thirdly, pressure from civil society. The chapter is in two parts. In the first part, I examine urban learning forums through the high-profile and increasingly mobile example of an urban forum in Porto Alegre, Brazil. In the second, I consider whether urban learning forums that take place in the context of *translocal* urban learning assemblages across the global North–South divide face particular challenges that accompany that very global categorization.

Learning the City: Knowledge and Translocal Assemblage, First Edition. Colin McFarlane.
© 2011 Colin McFarlane. Published 2011 by Blackwell Publishing Ltd.

This second part of the chapter begins with an example of an exchange between the Indian Alliance discussed in the previous chapter and a London-based homeless movement called Groundswell. As this and other examples show, urban learning forums that cut across the global North–South divide raise the spectre of this form of categorization. Categories like North and South, First and Third Worlds, developed and developing, can function to militate against the prospects of translocal urban learning forums. One implication of the resilience of these categories is that they are active imaginative barriers that can delimit the possibilities of learning between and through different urbanisms. In response to this, I use the concept of translation as a way of conceptualizing learning across difference. While the obstacle to the success of translocal urban learning in this example is a particular kind of perception of the global North–South divide, the more general point at stake here is that the success of translocal urban learning experiments depends upon a commitment to learning through translation – i.e. through difference rather than in spite of it – rather than simply through an attempt to learn through cities that appear to be 'similar'.

Uncertain Forums

At the centre of any discussion of urban learning forums is the notion of participation, an idea that has been vigorously attacked and disputed in critical scholarship as, at best, a distraction or naïve optimism, and at worst a form of political control of the marginalized. The scope to participate in the reassembling of urban life through learning forums has been severely curtailed within many contemporary cities, where conflictual politics and participatory difference is often shunned by a culture of managerialism, consent and consultation. As Ash Amin (2006: 1018) has written: '[T]he principle that urban public culture might be shaped through the free hand of a plural and equal citizenry has been compromised by an urbanism of differentiated rights and preordained expectations from the shared commons.' In contexts where urban marginals are increasingly 'tracked, gathered and shunted on' as threats to cities that are more and more dominated by corporate and consumer spaces, the very idea that urban life should allow for pluralism and dissent is sharply truncated (ibid.). As David Harvey (2008: 32) argued, this is an urban world that, far from promoting urban participation, is increasingly characterized by a 'neoliberal ethic of intense possessive individualism' where the 'political withdrawal from collective forms of action, becomes the template for human socialization'. Against this background, there seems little potential that urban futures in many cities might be formulated through what Amin (2006: 1018–19) called 'a politics of engagement rather than a politics of plan', unless civic states place 'confidence in the creative powers of disagreement and dissent, in the legitimacy that flows from popular involvement, and in the vitality thrown up by making the city available to all' (and see Mouffe 2000).

There is, of course, a wide-ranging set of literatures here that we could draw upon, from debates in urban studies or development studies on 'partnership', 'participation' and 'empowerment' through particular governance initiatives, to feminist and postcolonial debates on the performative force and possibilities of 'participation' (e.g. Ferguson 1994; Atkinson 1999; Cornwall 2000; Cooke and Kothari 2001; Lyons *et al.* 2001; Briggs and Sharp 2004; Williams 2004). These disparate literatures have variously criticized participatory initiatives as tools for governmentalizing subjects – for instance, through depoliticizing by rendering political questions as technical considerations – or as dealing with frivolous, short-termist concerns that only 'soften' neoliberalisms rather than address the causes of inequality and poverty. At one time an alternative discourse, participatory approaches have become mainstream, whether through the World Bank or through circulating discourses and strategies of urban participation such as those around participatory budgeting (Pieterse 1998, 2001; Mohan and Stokke 2000; Williams 2004; Sintomer *et al.* 2008). Following Cooke and Kothari's (2001) important elaboration of participation's disciplinary effects in the context of development initiatives, and their call for a rehabilitation of participation that is more attuned to its power relations, there have been instructive attempts to re-evaluate the possibilities of participation. For example, Glyn Williams (2004) highlighted examples of participation that actively enhance the political capabilities of the poor, including through the creation of new preferences, ways of seeing, new knowledges about rights and *de facto* rules of the game, and the politicization of previously ignored issues. Arguing for a political imaginary attuned to participation's claims to 'listen' and 'represent' – however problematic – and its implicit possibility of alterity, Williams (2004: 573) suggested that the progressive possibilities of participation are to be found 'within long-term struggles and reshaped political networks that link themselves to a discourse of rights and a fuller sense of citizenship'.

Participatory urban forums have their limits, to be sure, but they also represent one potential outlet for more socially just urban transformation. But how might urban learning collectives take shape? As Amin (2006) argued, this requires an extending and deepening of participation that encodes a heterotopic urban civic culture (Keith 2005) which allows dissent, difference and disagreement while confronting violence and encouraging expansive solidarities. Such a civic culture is not simply a rational realm of formal debate, but is an often passionate and embodied set of practices assembled and reassembled through interactions. A whole variety of experiential factors can shape the nature of participatory learning contexts involving different constituencies, to name just a few possibilities: fear over potential future displacement, anxiety over future development, vulnerability associated with living in poverty, hope for an alternative life, discomfort with sharing a meeting space with state officials or community organizers, changing atmospheres caused by a particular speech or intervention, frustration at the lack of change or the course of a meeting, anger over a particular comment, and so on. As John Forrester (2000: 151; cited in Bridge 2005) argued in his work on 'participatory rituals' in deliberative democracy:

> The analysis of learning through deliberative, participatory rituals suggests that we learn not only with our ears but with our eyes and hearts. We learn not only from surprising information that leads us to propose new hypothetical lines of action to test, but we learn from style and passion and allusion too. We learn to reframe our predictions and strategies, but we learn to develop new relationships and even senses of ourselves as well.

As Bridge (2005) showed in his discussion of overlapping traditions of deliberative democracy, pragmatist philosophy and social learning, participation is a dynamic social process. While these debates provide a useful focus on small group learning and interaction, they sometimes fail to appreciate the role of inequality in power relations in shaping group interactions and the sorts of learning that can and cannot take place. In contrast, argued Bridge (2005), strands of work in radical planning have sought to consider how planning might challenge existing power relations, for instance through community mobilizations or the work of radical planners working with communities, and facilitate more socially just transformations, for example through advocacy planning (e.g. Peattie 1994; Sandercock 1998) or feminist planning (e.g. Watson 1988; Leavitt 1994). Writing from the tradition of pragmatism and communicative reason, Bridge (2005) offered a conception of planning as argumentation and dissensus. Building upon a critique of deliberative planning and on a sympathetic but critical reading of Habermas, *planning as argumentation* recognizes that participants are not fully fledged debaters for whom communication is limited to linguistic action, but that gesture, feelings, and performance play often crucial roles in how people participate in urban planning meetings. Moreover, argumentation in this context is not only about who wins or whether a universal agreement has been achieved, but what people do to create change – for example, change to the atmosphere of a meeting or to the nature of potential outcomes: '[There is no] singular space of the public realm achieved via deliberation to consensus (in the Habermasian model). What we are left with are multiple spaces and multiple conceptions of space that transact as positions in argumentation' (Bridge 2005: 145). This form of planning seeks connection not in spite of but through conflict, diversity and difference. This description closely resonates with the conception of assemblage outlined initially in Chapter 1, that is, focusing on relations of history and potential, emphasizing performance, doing and events rather than resultant formations, and structured by unequal relations of knowledge, power and resource.

However, we are still left short here of a clear sense of the yardsticks through which we might appraise the nature and possibilities of urban learning collectives. In order to further develop the question of how urban learning collectives are produced and changed, I wish to turn to a book that examines this question in relation to science and technology controversies: Callon, Lascoumes and Barthe's (2009) *Acting in an Uncertain World*. Contemporary uncertainties around science and technology, from nuclear waste to genetic engineering, increasingly spill over into public debate (Whatmore 2006). Callon *et al.* attempt to consider how we might act in this uncertain world – what sorts of debate might proceed in what kind of contexts, and who might organize or facilitate these discussions? Knowledge controversies increasingly take place in what the authors call 'hybrid forums', public spaces in which a variety of groups discuss technical options, including experts, laypersons, politicians and technicians, examining heterogeneous issues and perspectives from ethics to the economic, politics to the technical (Callon *et al.* 2009: 18). These hybrid forums are 'an appropriate response to the uncertainties engendered by the technosciences – a response based on collective experimentation and learning' (Callon, *et al.* 2009: 18), within contexts of unequal power relations, defiant and dominant discourses, and questions over legitimation.

Uncertainty here refers to knowing that we do not know. It is an explicit recognition of the role of ignorance and error as part of the learning process. This distinguishes it from risk, which refers to an identifiable danger associated with a describable event or series of events. Risk involves the exploration of possible worlds, for instance in scenario planning, revealing the possibility of harmful events for particular groups. But uncertainty refers to contexts where we cannot anticipate the consequences of decisions, where we lack sufficient knowledge of the different options, where the description of possible worlds comes up against illegibility, and the behaviour of different people or their interactions remain enigmatic (Callon *et al.* 2009: 18–27). It refers to our inability to 'see' the dynamic totality of the learning assemblage. There is also an uncertainty about the dynamism of controversies: we often cannot know whether or how a debate will be resolved, and when that resolution might take place.

Callon *et al.* are not discussing urbanism here, but what is valuable for our purposes is the explicit recognition of the heterogeneity and uncertainty at work through forums as particular kinds of learning assemblages. This is not to say, however, that learning is necessarily a response to uncertainty. As Chapter 5 will show, learning can be a process of reinforcing existing positions and knowledges, or, as Chapter 2 argued, a means of negotiating different places and people within the city. Uncertainty is particularly important for learning forums that bring together heterogeneous groups and issues whose conclusions are purposively not predetermined. This uncertainty and heterogeneity, Callon *et al.* (2009: 28) argue, can enrich democracy through experimental formats: '[W]ith the hybrid forums in which they develop, they are powerful apparatuses for exploring and learning about possible worlds.' Forums operate as coordination devices for learning that negotiate different knowledges, situations, actors and interests, and which can generate new connections and problematizations, change the terms of debate, and bring to light certain possibilities for urban dwelling while suppressing others. They have the potential to themselves become spaces of tactical learning.

Learning here emerges from attempts to 'take into account' multiple voices, interests and expectations, i.e. the politics and uncertainties of translating and coordinating different domains, where 'translating' refers to the extent to which the knowledge of different individuals and groups is understood or misinterpreted by others involved in the forum, and where 'coordination' refers to the ways in which particular formats, discussions and individuals or groups bring together and structure multiple voices and knowledges, sometimes in exclusionary ways. It does not involve integrating other voices into pre-existing technical solutions, but goes further towards collective engagement amongst the ever-present threat of claims to authoritative discourses. It is, to be sure, an optimistic reading of learning that potentially 'allows laypersons to enter into the scientific and technical content of projects in order to propose solutions, and it leads the promoters to redefine their projects and to explore new lines of research able to integrate demands they had never considered' (Callon *et al.* 2009: 33). This means, potentially, the production of new-shared knowledge, ways of seeing and acting, emerging from two springs. Firstly, and within contexts of asymmetrical power, the unusual nature of confrontation between specialists and laypersons can force new lines of inquiry to be taken into account. At stake here is learning not just different forms of knowledge, but different ways of seeing issues – an education of attention. Secondly, there is – again, optimistically – a kind of attuned ear at work as different groups question their perception of each other through interacting, opening the way for compromises and alliances.

All of this occurs not through the simple addition of information or the aggregation of points of view, but through trial and error, argument and conflict, inclusions and exclusions, and constructing and deconstructing, and the stakes are often high for particular groups. Hybrid forums are dialogical spaces involving the double movement of laypersons and specialists from historically segregated spaces, and can become an 'apparatus of education' (Callon *et al.* 2009: 35). Callon *et al.* (2009: 153–90) identify three organizational criteria that facilitate learning forums: *intensity, openness* and *quality,* yardsticks for appraising the extent of dialogism, i.e. the emergence of new possibilities, collectives and identities. Intensity refers to the formulation of the forum: how early on are 'laypersons', to use their term, involved in the exploration, and how intense is the concern around the composition of the collective? Openness refers to the degree of diversity of the groups consulted and their degree of independence *vis-à-vis* established action groups, as well as the extent to which spokespersons are allowed to speak on behalf of their constituencies (thereby potentially silencing marginal voices). Quality refers to the *seriousness* of voice – whether people are actually allowed to deploy their arguments and claims – and the *continuity* of voice, i.e. whether interventions are sporadic or lead to focused conversations. These three criteria are then set against three procedures: equality of conditions of access to debates; transparency and traceability of debates, i.e. procedures to ensure that voices have been recorded; and clarity of rules in organizing debates, ensuring that voices are heard and that dominant groups do not structure proceedings.

In other words, intensity, openness and quality are measures of learning as coordination – i.e. the extent to which the coordination of different groups and knowledges facilitates genuinely inclusionary, participatory discussion where different ideas are heard and translated – that need constant attention as they are played out in practice. While there are distinct echoes here of Jürgen Habermas' (1962) conception of the public sphere in the focus on debate, discussion and exchange, Callon *et al.* (2009: 263) dismiss this connection by arguing that – and this is a charge they also level at John Rawls and Hannah Arendt – Habermas imagines individuals stripped of their identities, attachments, anxieties, ambivalences, singularities and differences, and instead absorbed by nothing but their will to communicate. Callon *et al.*'s claim here is that it is precisely the 'attached and cluttered' nature of the individual that allows people to reach provisional understandings with one another through forums. The three useful criteria of intensity, openness and quality can be applied, with due care, to urban learning forums which seek to bring together different people, knowledge and perspectives in the context of urban development. In the next section, I consider examples of the potential of forums to facilitate learning in specific relation to the city.

Dialogic Urban Forums

Some of the most promising experiments with urban learning forums have taken place in Brazil. In post-dictatorship urban Brazil, there has been a tradition of participatory democracy that has been particularly associated with its vibrant social movements and the *Partido dos Trabalhadores* (PT, the Workers Party), especially in its high-profile successes in Porto Alegre. Before the PT assumed power in many Brazilian cities from the early 1990s, movements such as Rio's *Movimento de Associacoes de Barrio* (Movement of Neighbourhood Associations) experimented with establishing neighbourhood discussion forums and sought to make demands on the state. National movements like the *Cost of Living Movement* operated in similar ways, and some Mayors under the military dictatorship sought methods of establishing participation. With the collapse of the military dictatorship and the PT's growth and eventual success in national elections came decentralization of power to local levels, especially through the 1988 Constitution (and, simultaneously, more pressure on municipalities to deliver services and compete with other cities for capital). Curitiba, for example, established well-known citizen participatory forums for transportation and the environment from 1989, with notable success (Baiocchi 2003). The key urban success area to date for the PT has been in participatory budgeting, especially in Porto Alegre where there has been mass participation, elements of redistribution, and a balanced budget. Citizens decide and deliberate upon a variety of municipal policies, the cornerstone of which is the much-publicized *Orçamento Participativo* (Participatory Budget), a neighbourhood-based set of deliberative forums on the city's budget priorities.

As an experiment in participatory urban forums, Porto Alegre stands apart in Latin America for its scope and sustained commitment, and is one of the most successful cases of municipal socialism on the continent (Abers 2000; Baiocchi 2001; Sintomer *et al.* 2008; Goldfrank and Schrank 2009). When the PT assumed power in Porto Alegre in 1989, there was already a thriving debate about participatory planning in the city. The Union of Neighborhood Associations of Porto Alegre, for instance, had in its 1985 congress called for a participatory structure involving the municipal budget, and the PT was supportive. In the four consecutive mayoral terms in which the PT governed Porto Alegre (1989–2004), the PT was able to build on its existing largely middle-class base by drawing in an extensive working-class constituency through courting and eventually winning supporters of the Democratic Labour Party (PDT) (although later, in 2004, the middle class deserted the PT and cost the party the municipal elections; Goldfrank and Schrank 2009).

With a third of the population living in favelas, the administrators responded to the challenge of involving the poorest in participatory planning by developing regional *orçamento participativo* assemblies in each of the city's 16 districts. Learning through participatory budgeting is coordinated by a strict and clear set of procedures for organization, representation and participation. Participation is two-tiered, involving both individuals and community organizations. Meetings begin in March, when delegates are elected to represent specific neighbourhoods, and the previous year's projects and budget are discussed. The number of regional delegates elected per neighbourhood to participate in subsequent deliberations is decided based upon attendance. In the months that follow, these regional areas meet to discuss local and city-wide priorities, and to examine thematic areas like health or education. Finally, regional delegates come together at the regional plenary to discuss local priorities and to vote to elect councillors – whose term is limited to two years – to serve on the Municipal Council of the Budget (Baiocchi 2001; Sintomer *et al.* 2008).

This council is tasked with reconciling the demands from each region with available resources, and proposing and approving a municipal budget in conjunction with members of the civic administration. Its 42 members and representatives of planning agencies meet biweekly for several months in this process. These forums – which have led to higher levels of local participation every year – function as a space for local demands and problems to be aired, for information to be divulged about the functioning of government, and as a regular meeting place for activists. In addition, as Baiocchi (2001, 2003) argued, the ongoing year-on-year cycle of debate and participation in local areas and city-wide allows space for people to learn from mistakes. He argued based on his research in three regions in the city that the key emphasis of local meetings is not in making decisions about budgets, but in learning about the technical criteria involved in budgeting. The process is a combination of participatory democracy (regular regional debates and learning initiatives) and representative democracy (e.g. in the Municipal Council of the Budget). The local assemblies can change the rules on participation, voting systems, and topics for debate yearly if they choose to. Baiocchi's (2001) research shows that once people become involved in participatory budgeting, they are more likely to take part in participatory forums in other specific sectors, including education, health, infrastructure services, sports facilities, and so on (see Abers 2000).

These efforts have not, of course, been without their critics. As Biaocchi (2001) argued drawing on Bourdieu (1991), a central criticism of these sorts of participatory forums is that they can reproduce class hierarchies, giving increased influence to local élites, and that they reproduce hierarchies of political competence of 'experts' against non-experts. In this critique of the fiction of 'linguistic communism' (Bourdieu 1991), the competence to speak embodies difference and inequality – a privileged class habitus structures the technical ability to speak and the standing to make certain statements. However, Biaocchi argued that in Porto Alegre the poor actually participate more than the élites partly because the notion of 'need' is a primary motivator for the participatory forums *tout court*, and partly because of the atmosphere of informal expectation that marginal voices should be heard. In addition, the average participant in regional meetings is of lower formal economic and educational standing than the average citizen of Porto Alegre, and there is little if any evidence to suggest that those with higher levels of educational attainment have more chance of being elected – indeed, if anything the evidence suggests that the reverse is the case.

More importantly, in contrast to what we might expect to see if more powerful groups were manipulating the process, the vast majority of investment has gone to poorer areas of the city and has affected poorer citizens. In the years between 1992–1995 the housing department offered housing assistance to 28,862 families, against 1714 for the comparable period of 1986–1988, while the number of functioning public municipal schools jumped to 86 against 29 in 1988 (Biaocchi 2001). Sintomer *et al.* (2008: 166, 167) wrote:

> [Participatory budgeting] has provided for a reversal of priorities: primary health care was set up in the living areas of the poor, the number of schools and nursery schools was extended, and in the meantime the streets were asphalted and most of the households have access to water supply and waste water systems ... [revised budgeting formulae ensure] that districts with a deficient infrastructure receive more funds that areas with a high quality of life.

The extent and regularity of the municipal bus company service was massively increased, thousands of families received public housing or land titles, and a large network of daycare and health clinics was established (Goldfrank and Schrank 2009). All of this was paid for through progressive taxation and a crackdown on tax evasion (ibid.). However, none of this means, of course, that the views of the poor and the better-off register equally, and there is no reason to believe that participatory forums somehow undo existing inequalities. In addition, while there is no evidence of gender imbalances in terms of numbers at the meetings, there is some evidence that woman are less likely to speak than, or indeed be listened to by, men (Biaocchi, 2001, 2003).

Learning is a central part of the Porto Alegre forums. Biaocchi (2001) quoted one participant:

> I had to learn about the process as the meetings took place. The first time I participated I was unsure, because there were persons there with college degrees, and we don't have it, so we had to wait for the others to suggest an idea first, and then enter the discussion. And there were things from city hall in the technical areas, we used to 'float'. But with time we started to learn.

Over time people translated 'technical' knowledges in ways that facilitated their understanding and participation. People learn, over the course of a year, about each other's priorities and about the technicalities of budgeting, policy development, and the operations of different public services. Indeed, Sintomer *et al.* (2008) argued that the establishment of transparent rules has largely overcome the clientelistic structures of urban planning that existed previously. City officials attend local weekly meetings, taking part as facilitators whose role is to foster rather than interfere with discussions. The forum exceeds the confines of the state while working, depending on context, in collaboration or opposition to it. For instance, the number of neighbourhood associations has grown exponentially since the participatory budgeting process began. Often, these associations take urban learning as their focus. One example that Biaocchi (2001) highlighted took as its first aim the need 'to obtain and share information about the municipal administration' – echoing the work of the Federation of Tenants Association discussed in Chapter 2.

Many of these associations have taken on a 'problem-solving' ethos of learning-through-dialogue in how they address specific urban services and institutions. These associations have altered the relations between the state and civil society in Porto Alegre. Rather than these relations being characterized by citizens waiting, often individually, for long periods of time in the offices of officials in order to discuss a service, now state departments and neighbourhood groups are involved in regular dialogue about the possibilities of existing and emerging services. Moreover, neighbourhood groups are increasingly brought into contact with one another through the participatory process, constituting the ongoing formation of translocal urban learning assemblages that exchange ideas, negotiating strategies and technical expertise across different neighbourhoods in the city. For instance, one resident who went through the participatory budgeting process described to Biaocchi (2001) a new translocal urban imaginary through learning:

> As delegate and councillor you learn about the region, meet new persons, become a person who has to respond not only to your association, but also to the region as a whole and the city as a whole. I participated in the two congresses to decide the *Plano Diretor* [municipal planning priorities] and since I have worried about the city as a whole. After a year, I learned not to look only at the region, but that you have to look at the city as a whole.

The success of urban participatory forums in Porto Alegre is, to be sure, partly a function of the city's relative wealth compared with other cities in Brazil. But it is about more than this. The municipal government, in the form of the PT, was committed to experimenting with participatory democracy and – crucially – was aware that this had to mean allowing autonomous civil society debates and institutions to flourish. Although these conditions are not necessarily reproduced throughout urban Brazil, there is none the less widespread debate around the possibilities of urban participation in the country. For example, at the national level, where the PT holds power, the 2001 City Statute provided for participatory budgeting and imagines 'citizens who are active, organised, and well informed' (Caldeira and Holston 2005: 406). The Statute shifts the urban state to role of manager, managing public, private, and community interests. This allows, of course, for not just a deepening of democracy (e.g. in the participatory budgets), but simultaneously an exclusionary neoliberal agenda of privatizing infrastructure to generate new funds (e.g. Silva 2000).

With assuming power, PT administrators sought to balance workers interests (e.g. pay rises) with fiscal controls and the other investment needs of the city. In São Paulo, for example, Luiza Erundina's PT administration, which included Paulo Freire, lasted one term in the early 1990s during which it made substantial improvements in housing, infrastructure in favelas, education and health, but also increased bus fares and failed to handle disputes within the PT and across the city effectively. Efforts at participation of citizens substantially failed to get going, and many in the poorer neighbourhoods who had voted for the PT felt alienated. PT lost to Paulo Maluf's conservative party who then governed for two terms, during which participatory budgeting was forgotten and the municipal budget was significantly overspent (Couto 2003). The PT won power again in São Paulo in 2000 with Marta Suplicy's leadership, a Mayor from an upper middle-class background with few previous PT links. In contrast to Erundina, who tried to lead without coalitions and operated in a split between the PT party at large and the PT party in the Mayor's office, Suplicy sought to build coalitions from the start, and her administration was less held back by internal divisions. If, for Couto (2003), the PT in São Paulo has subsequently governed more successfully through coalitions, one implication has been the marginalization of participation due to the distractions of playing a particular kind of political game. This realpolitik has limited the conditions of possibility for the development of urban participatory forums in the city. More generally, PT activists are faced with the additional challenge of ensuring participation in areas where there is a relatively weak civil society.

Significantly, the success of the forums is a product of a commitment to a particular ethic of learning as coordination, i.e. to long-term genuine participatory discussion amongst different groups that is able to be carried through in the sense that it has decision-making capacities in areas such as redistribution of spending. This coordination can be assessed using the intensity-openness-quality schema. The *intensity* of the Porto Alegre participatory forums is reflected in the early involvement of different constituencies in the urban planning process and in the explicit concern that marginalized groups be involved, while the *openness* of the process is found in the combination of participatory and representative formats, allowing people to monitor what those elected on their behalf say and to depose them if necessary. The *quality* of the process resides in the continuity through which participation is enacted, maintained and evidenced in the redistributive budgets that result. However, as useful as the intensity-openness-quality criteria are, we need to be careful not to underplay the importance of the political context within which urban learning forums take place. As the comparative experience of Porto Alegre and São Paulo demonstrates, different political contexts enable or hinder the possibilities of urban learning forums to flourish. Indeed, as Abers (2000) argued, it was a rare 'window of opportunity' in Porto Alegre that allowed participatory forums to succeed as significantly as they did: a combination of a newly elected workers party committed to people-oriented development, and an active and diverse grassroots movement determined to involve ordinary people in the post-dictatorship planning of the city.

The examples from Brazil raise the question of what sorts of criteria and procedures might function as practical tools for implementing urban learning forums elsewhere. In their study, Callon et al. (2009) highlight several potential formats: *focus groups*, which the authors view as useful in identifying priorities, but which are episodic and generally do not lead to changing relations between experts and laypersons; *public inquiries*, which Callon et al. claim succeed only where there is genuine commitment to involve the public; *consensus conferences*, new expert–layperson forums that focus on particular issues by raising awareness, stimulating debate, and leading to the production of citizen reports – a meaningful start, but often not a sustained collaboration; and *citizens' panels and juries*, which often privilege local points of view but which are rarely about dialogue. If none of these procedures are themselves satisfactory for dialogic democracy, specific procedures will be more or less relevant for particular issues and at particular times and spaces. These four formats are clearly suited to the sorts of science and technology controversies that Callon et al. (2009) write about, and urban experiments such as participatory budgeting offer an alternative set of procedures. However, there is no reason why, for instance, local focus groups could not be used to learn about and identify priorities in relation to, say, the construction of community facilities in a poor neighbourhood, neither is there any reason to ignore the possibility of urban consensus conferences whereby citizen reports would be produced on issues as different as budget priorities, investment decisions, daycare, or the locating of urban dumping grounds, perhaps generating media attention. Urban learning forums should, then, be contingent on the issues at hand, and driven by an experimental ethos that is committed to collective learning. Moreover, rather than an appendage or bureaucratic procedure within urban planning, if they are to succeed they should be central to the very conception and nature of urban planning itself.

What should emerge from these urban learning forums? By definition, we cannot know. However, we might speculate on the nature of those outputs. For Callon et al., the idea that forums should lead to finalized decisions is contrary to their spirit; they should be open to new information, voices, events and revisions. Callon et al. (2009: 191) talk not of decisions but of 'measured action' – an apt descriptor for Porto Alegre's urban democracy. Decisions in this context are less clear-cut, and more a series of rendezvous involving a changing urban learning assemblage of actors with diverse responsibilities, and which are sometimes subject to new formulations.

Urban forums can be seen, then, as particular forms of urban learning assemblage – centralized, organized but also dynamic experiments that can 'generate turnarounds in opinions and encourage the review of the best-established agreements' (ibid. 245). These agreements need not be watered-down consensus. In fact, and as the history of radical planning has demonstrated, at best these agreements are transformations that facilitate more socially just solutions. In critical urban scholarship, the most radical antecedent and symbol of revolutionary gathering of the poor and marginalized remains, of course, the Paris Commune, celebrated by Debord and Lefebvre as a festival against alienation and oppression through which, in Lefebvre's words, 'a scattered and divided city became a community of action' (1965, reprinted in McDonough 2009: 174). Urban learning forums belong not in the governmentalizing tradition of state consultation, but in the radical and hopeful tradition of rethinking urbanism through argumentation, contestation and reformation.

One important question that we have not addressed so far in this discussion of urban learning forums is that if these forums are distanciated – i.e. if they are based on exchanges between groups living in multiple different places, sometimes different countries – do we need to have a different kind of forum in mind when we consider learning that brings people together across far-flung sites? In particular, how might we envisage urban learning forums that cut across (stubbornly persistent) global North–South divides? Translocal urban learning forums are becoming increasingly important for urban policy, planning and activism (see Chapters 3 and 5), serving as a crucial reminder that participation and collective learning need not be spatially conceived simply as 'local' or territorial. The next section examines these questions through particular case studies.

Translocalism and Translation

The idea of participatory budgeting forums has travelled extensively from Porto Alegre. More than 1,000 municipalities in South America have adopted some variant of it since the early 1990s, and by 2008 more than 100 European cities were developing forms of participatory budgeting (Sintomer *et al*. 2008: 164). Sintomer *et al*. (2008) conducted comparative research on participatory budgeting in more than 20 cities, and showed how the method has moved through quite distinct circuits and been translated into very different forms. For instance, one means through which the method has travelled has been the World Social Forums, which in 2001–3 and 2005 were held in Porto Alegre, and participatory budgeting has become a 'symbol' of participatory democracy amongst certain strands of the anti-neoliberal globalization movement. To take a different instance, both the World Bank and UN Habitat have described Porto Alegre's example as one of 'best practice' in relation to urban policy (e.g. Cabannes 2004; UNDP 2001). In February 2008, the World Conference on Development of Cities was held in Porto Alegre, involving 7,000 mayors and city councillors, academics and experts, community leaders, business people and social workers. The conference focused on the diversity of social transformation initiatives that have been emerging globally through four themes: (1) the right to the city; (2) governance and democracy in cities; (3) local development in cities; and (4) sustainability and the network-city. The outputs from these kinds of large international events are difficult to gauge, and often reside more in developing broad areas of agreement, for instance in relation to the importance of participation of diverse groups in urban planning. Important differences, of course, exist amongst the various agencies involved, but the sheer breadth of the different actors involved in the 2008 conference provides an indication of the growing translocal popularity of participatory urban forums: the Porto Alegre City Council; Brazil's Ministry of Cities; City Hall of Rome; Rio Grande do Sul State Government; the National Confederation of Cities; the Federation of Latin American Cities, Municipalities, and Associations; the UN's United Cities Local Government organization; the UN Educational, Scientific, and Cultural Organization; UN Habitat; the Inter-American Development Bank; the World Bank; the International Observatory of Participatory Democracy; the Committee on Social Inclusion and Participative Democracy; the International Center for Urban Management; and Cities Alliance (Cities Alliance 2008).

In other cases, particular urban policy-makers adopt the example from Porto Alegre either through visiting the city or reading about it. In Bradford, UK, for example, participatory budgeting was used to promote investment in marginalized areas, although the sustained involvement of local communities in the urban planning structure was far less developed than in Porto Alegre. In addition, it is not as if Porto Alegre's example is the only form of participatory urbanism that is travelling. We might also consider the additional influence of trends in New Public Management emphasizing consultation rather than participation (Desai and Imrie 1998). For instance, Sintomer *et al.* (2008: 173) wrote: 'In Germany, [New Public Management] was imported from the New Zealand city of Christchurch, and the Brazilian experiments had an impact only later on, leading to the emergence of mixed models.' Urban learning forums are assembled by translating ideas or models from elsewhere, prompting important questions and challenges. The success of travelling forms of participatory budgeting depends upon political commitment from the state, pressure from civil society groups, and existing political traditions of political democracy and participation (Sintomer *et al.* 2008). In addition, questions emerge specifically from the translocal element of urban learning forums.

For example, translocal urban learning forums that cut across the global North–South divide face particular challenges, and can be constrained by stereotypical categorizations that can limit those opportunities. Geographical categorizations such as global North and South play an important role in affecting how learning around urbanism occurs. To take an example related to Slum/Shack Dwellers International (see Chapter 3), in one exchange between an NGO working with homelessness in London – Groundswell – and the Indian Alliance NGO SPARC based in Mumbai, it was the perception of the UK group by SPARC leaders that constituted one of the main limits to this particular learning experiment. While this short-lived translocal forum involved policy-makers and donors, it was different from the Porto Alegre example in that it was more dominated by exchange and solidarity between civil society groups and activists, but what is important about this example is the way in which perceptions around the global North–South divide affected the possibilities for learning through translation. This is a challenge that is potentially faced by any translocal urban learning forum that crosses North and South.

The exchanges to Mumbai and London involved activists from both cities as well as a small group of policy-makers and donors. SPARC had chosen to get involved in the UK exchange, funded by the British donor Homeless International (2001), because of what one SPARC leader described as 'romanticisation with the North'. There was a belief by some SPARC leaders that they had little to learn from Groundswell because of their perception that Groundswell – as an NGO from the global North – was dependent on state welfare and focused on receiving state financial assistance rather than developing their own solutions and funding mechanisms. One SPARC official invoked a broad distinction between First and Third Worlds. She argued that there is little potential to learn between Indian and UK groups because of the 'socio-political culture' in the UK. She pointed to what she viewed as a general 'lack of community' in British cities, adding that the British were too 'individualistic', and extended this view to NGOs she had encountered in Japan. Such comments imply a polarizing of Northern (here including Japan) and Southern urbanism, where the former stands for individualism and the latter an active collectivist focus.

This perception played a role in militating against the opportunities to learn between these two groups. The opportunities for SPARC to learn from the exchange were limited from the start. The suggestion from SPARC was that Groundswell and the UK urban voluntary sector more generally was passive because of a reliance on subsidy and therefore too different to be able to offer SPARC and its members any opportunities to learn. For example, SPARC leader Sheela Patel (1997) wrote in one report – *From the Slums of Bombay to the Housing Estates of Britain* – following a visit to UK urban voluntary groups, that participation in the UK is reduced to state consultation. Her argument was that this consultation separates people from officials and boxes them into groups that should be consulted: 'It was as though participation was the right to choose the tiles, the cooker and the door frame from a selection made by the architect and on display in the model room' (Patel 1997: 25). This she contrasted with her own view of participation as an intrinsic part of the urban development process, a process that includes a wide variety of actors and is driven by the poor as self-managers:

> In the world we live in today, we are all inter-dependent to a very large degree. We depend on each other to improve the quality of our lives. Yet each of us has the right to determine what makes us happy, what creates conditions to actualise our aspirations, and develop skills and the capacity to present to others whose interdependency with ourselves requires us to come to a consensus about how things should go forward. That is the process of participation. (Patel 1997: 25)

On the other hand, Groundswell reported that they learned a great deal from SPARC. They gave several examples: the use of local 'horizontal exchanges' as a learning and 'capacity-building' tool, the value of enumerating the 'hidden homeless' within cities as a basis for state negotiation, and on the relationship between Groundswell as an NGO and the community-based organizations it is linked with in the UK. The then-coordinator of Groundswell, Toby Blume, argued that Groundswell has drawn much of its 'insight and inspiration' from the Alliance (Blume 2001). Members of the groups visited the Indian Alliance in Mumbai in January 2001, and following this visit Groundswell initiated a programme of exchanges in the UK, 'having recognised exchange as a powerful tool in strengthening our network' (Blume 2001). According to Blume, Groundswell 'came across the idea of exchanges when we came into contact with groups of homeless people from India and South Africa' in London in January, 2000 (Groundswell 2001a). For Blume, participating in and documenting exchanges, along with 'Speakouts' – where various groups come together annually from around the UK to engage policy-makers – and communications through e-mail, telephone calls, or the Groundswell newsletter, results in a 'recording of the learning process' and the expansion and refinement of the 'tool box' (Blume 2001).

Blume returned from India as an 'exchange evangelist' and immediately set out to start the process of 'experimental learning' (Blume 2001). Groundswell began a round of exchanges amongst groups of low-income tenant community groups, homeless groups, drug and alcohol addiction groups, and policy-makers. These included visits to Bristol, Glasgow, Birmingham, Nottingham and London, all in 2001. Blume said Groundswell had learned 'about the validity of how we go about galvanising action in the UK [from the exchanges with the Indian Alliance]'. For the Groundswell members, there was an important role for solidarity based around the common experiences of urban social injustice in forming relations between the UK homeless visitors and India's urban poor. A Groundswell (2001a) diary report written following a Groundswell–Alliance exchange in Mumbai argued that while conditions in India were radically different from those in the UK, 'for the UK group, the work of the [SPARC affiliates] NSDF and Mahila Milan had been extremely inspiring, and it was surprising and exciting that they shared many common experiences. Although the problems and challenges of homelessness and poverty might be different, the process for involving homeless people in creating the solutions could be very similar indeed.' Other people working with Groundswell who met with Indian visitors at an event in January 2000 in London commented on what they viewed as a potentially productive learning relationship:

> We have to join together as the people who are actually living the problem, not the people coming in and telling us what the solution is. If we do that in this country and actually form a federation very similar to what they have in the South then we can actually federate with them around the world and have a unified voice. (Newton, in Groundswell 2001b)

SPARC's conception of learning, at least in this temporary North–South translocal urban learning forum, is closely related to its perception of Groundswell and its Northern context, and stands in contrast to how the NGO and SDI more generally purport to conceive learning as exploratory, creative and open-ended (Chapter 3). This particular translocal urban learning forum was heterogeneous in that it brought different people and concerns together, but the intensity, openness and quality of the experience were curtailed from the start, even if it proved useful for Groundswell. If SPARC had conceived learning with Groundswell as indirect, open and unpredictable, that is, as a process of translation through difference, rather than as restricted to the closed direct transfer of knowledge or experience (a view that required for SPARC some measure of similarity in socio-economic context), then a more productive and sustained engagement may have been possible. The conceptualization of learning as necessarily direct dismisses possibilities because of (real or perceived) differences, and is testament to the ongoing role of divisions of global North and South as imaginative barriers that translate spaces into a comparative frame based on similarity, even in a context where the participants had opted to participate in the exchange (Edwards and Gaventa 2001). Exploring translocal learning forums among civil society groups often involves addressing these sorts of myths and geographical stereotypes.

Indeed, in other cases, exchanges have served to *disrupt* stereotypical perceptions of global North and South. Writing about the experiences of people from the US in visits to India and Mexico, Gaventa (1999: 35) pointed to the 'amazement at the knowledge, commitment and sophistication' participants found – 'a reality that did not fit with their received images of "backward" people.' He continued:

> Moreover, they often gained inspiration from the commitment which they saw ... 'By getting rid of our myths, we create the desire to learn more. Understanding that we have been taught wrong and then looking at the problems and consequences of that misteaching creates enormous openings. It's like turning a rock into a piece of clay that wants to be malleable by choice.' (Gaventa 1999: 35)

What these experiences hint at is the possibility that learning can occur not just in spite of differences, but *through* them. This notion of learning can present new opportunities for translocal urban learning forums, and points to the possibilities created by conceiving learning through translation rather than direct transfer between ostensibly 'similar' contexts. One instructive and provocative attempt to explore the possibilities of learning between North and South is a special issue of an Institute of Development Studies (IDS) Bulletin of 1998 entitled *Poverty and Social Exclusion in North and South*, edited by de Haan and Maxwell.

De Haan and Maxwell (1998: 5) contended that it would be 'foolish to deny the possibility of learning across geographical boundaries', and made a series of interventions on different themes of development. The theme issue highlighted a number of specific areas where connections can be made, including the nature of 'active labour market policies' designed to help people find work (Robinson 1998), the nature of participation in development programmes (Gaventa 1998), alternative routes to the reform of social welfare (Evans 1998), and the value of food security analysis (Dowler 1998). De Haan and Maxwell (1998: 7) made some further suggestions: 'What, for example, can we learn in the North from the successes with employment guarantee schemes in India or Botswana?' They made suggestions for joint research projects on specific themes: '[S]mall-scale credit, participation and participatory methods, social policy, food policy, and public works; and, indeed, in the meaning and measurement of poverty and social exclusion' (de Haan and Maxwell 1998: 8).

In their editorial, de Haan and Maxwell (1998: 7) argued that even if knowledge cannot travel directly, attempts to learn between North and South could still be 'fruitful'. In his contribution, Maxwell (1998: 24) argued that 'the point here is not to pretend that analysis and policy from one country can be read off directly from another, even within broad groupings of North and South. It is simply to demonstrate that opportunities are missed to compare and contrast.' However, elsewhere in the issue de Haan and Maxwell (1998: 7) inserted a caveat: 'Despite growing heterogeneity among developing countries and some signs of convergence between the North and parts of the South ... the particularities of place and history remain important, so that lessons can rarely be transferred directly.' This is an important point, but perhaps a more useful way to conceptualize this in relation to translocal learning forums is to emphasize that *because* the particularities of place and history are important, learning can occur but usually indirectly. This requires the understanding that knowledge and ideas can change in new circumstances, and that learning can occur in creative, indirect ways. For instance, specific strategies in the 'South', like public works, food policy or urban participatory budgeting, may appear to offer little opportunity for learning in the North if the approach is to ask whether the strategies can be transferred directly. They may offer more, however, if the approach is to engage in debate around these strategies without a rigid predetermined notion of how they may be useful. More general debates about urbanism, such as those concerned with the meaning and measurement of terms like poverty, development, urban economy or culture, also offer a basis through which learning through translation may occur.

Rather than focusing simply on the question of whether knowledge remains the same or not, translation focuses attention on the multiple forms and effects of knowledge. Translation embodies a sense of creative possibility that does not reduce learning to direct transfer. In this context, we might usefully invoke Said's (1984) development of 'travelling theory' as a basis for conceptualizing the possibilities of learning in translocal urban forums.[1] Said argued against the tendency to seek to apply theories wholesale or to dismiss them as completely irrelevant. He argued that the use of theory need not be reduced to this binary construction, and regretted that much intellectual work is caught up in what he viewed at the time as an anxiety and/or criticism over the question of misinterpretation:

> It implies, first of all, that the only possible alternative to slavish copying is creative misreading and that no intermediate possibility exists. Second, when it is elevated to a general principle ... [it] is fundamentally an abrogation of the critic's responsibility. ... Quite the contrary, it seems to me possible to judge misreadings (as they occur) as part of a historical transfer of ideas and theories from one setting to another. (Said 1984: 237)

This notion of 'misreading' focuses attention on the importance of change and the possibility of using what is witnessed, experienced or read about in one place in a way that need not be about trying to copy and directly apply it in another. The concept addresses a politics of replication by emphasizing the importance of creativity and local relevance. An emphasis on indirect learning rallies against the dismissal of, for instance, a place, knowledge or

---

[1] Said (1984) gave two examples of 'travelling theory': one the travelling of a theory from revolutionary Budapest to Paris, and the other of Foucault's theory of power; and argued for the importance of thinking cautiously over whether theories from elsewhere are relevant and how they can be changed for a new setting.

an idea as wholly irrelevant, and draws attention to the creative and uncertain possibilities of misreadings. Concepts like translation or perspectives like travelling theory open possibilities for indirect learning and assert transformation over transfer. As the example of the SPARC-Groundswell exchange demonstrates, translocal urban learning forums that cut across the North–South divide must contend with the ongoing representational power of those global geographical categories. If such translocal urban laboratories are attentive to their intensity, openness and quality, and if that attention is reciprocated by political actors, then translocal urban learning forums may become increasingly commonplace and influential. The seeds of such formations are, of course, constantly developing, often with distinct temporalities – short-lived or long term – and spatialities that fold different sites, organizations and formats, from conferences and workshops to exchanges and Internet forums.

For example, Cities Alliance is a loose coalition of governmental and non-governmental organizations that explores urban development strategies through learning from different actors, from movements like SDI to think-tanks like the International Institute for Environment and Development and international agencies like UN Habitat. They have become increasingly involved in translocal urban learning forums in different parts of the world. For instance, in 2008 they were involved in the São Paulo International Policy Dialogue among South–South megacities to share lessons on the 'Challenges of Slum Upgrading', with São Paulo's experience as the case study, which included groups of activists and policy-makers from India, Nigeria, Egypt and the Philippines. If these translocal urban learning forums are becoming more frequent, their connections are increasingly folded into website presences that assemble information and coordinate different actors. SDI, for instance, is similarly aware of the importance of the Internet in marketing its work and facilitating translocal urban learning forums. The Internet provides a quick means of disseminating information, staying in touch, and maintaining links for those literate and with computer access. For example, the Indian Alliance website was at the time of writing being overhauled to act as a data storage source for activists (www.sparcindia.org). Websites should not necessarily be viewed as separate to translocal urban learning forums that take place 'on the ground'. They are often actively folded into the emergence and maintenance of such forums (Juris 2008). In the case of the Alliance, for example, the website functions alongside workshops and conferences, such as the Mahila Milan (Woman Together) workshop on enumerations in Maharashtra held in February 2009, or the National Slum Dwellers Federation (NSDF) national conference on urban development, which brought together different NSDF groups to share experiences in Bangalore in April 2009. Different formats can serve to support one another, playing a role in the intensity, openness and quality of urban learning forums as they emerge and dissipate.

As the examples of participatory budgeting from Brazil demonstrated, if translocal urban learning forums are to influence policy and planning, then the commitment of states is crucial. For instance, in June 2009 SDI federations from Malawi, Zimbabwe and Zambia visited the SDI group in Namibia to celebrate its 10-year anniversary by joining in the construction of 48 brick houses in one neighbourhood in Gobabis. A reflection of the level of state support the Namibian SDI group has enjoyed was evidenced in the fact that joining in on the construction was the Prime Minister and various other government officials, including the Mayor. The Gobabis Municipality received UN Habitat's best practice award in 2007. Translocal assemblages like these are translocal in that the politics and practices of learning are in part a product of ongoing forums across southern Africa and beyond, but their degree of intensity, openness and quality are dependent upon local state commitment to the participation of the urban poor. For translocal urban learning forums to succeed, there needs to be a commitment on the part of the local urban state involved to cede decision-making power to the forum – to allow, for instance, the sort of redistributive budgeting that we see in Porto Alegre. But translocal urban learning forums do not depend upon state commitment alone. There are at least two other requirements. Firstly, as the Porto Alegre experience has shown, there is a need for an active civil society with a commitment to justice for the poor and marginalized that will help to ensure a strong participatory ethos from the start. Secondly, the possibilities of learning are dependent upon an openness to learning through difference, that is, to learning as translation rather than reducing the possibilities of learning to direct knowledge transfer. An important geographical challenge here lies in negotiating persistent stereotypes of global North and South.

Conclusion

This chapter has attempted to examine the sorts of environments that might facilitate more progressive possibilities of urban learning. Urban learning forums entail the possibility of different actors and knowledges within the city coming together to participate, in the context of unequal power relations, in collective learning. If managed carefully to facilitate sustained intensity, openness and quality, these learning forms take urban planning in uncertain directions and increase the possibility of more socially just urbanism. In urban learning forums such as those in the Porto Alegre participatory budgeting programmes, which are examples of tactical learning, there is evidence that forums are integrated into urban planning, not as an appendage, but as the very form of planning itself. In this kind of learning experiment, there is the potential – provided that the state cedes decision-making power to the forum – for the emergence of a different kind of city.

As unlikely as these sorts of participatory learning experiments may often seem, we should not allow ourselves, in my view, to cede the ground to power by becoming overly cynical about the possibilities of collaborative and dialogic urban learning. However, if urban learning forums are to be active constituents in reassembling the city, there must be commitment year-on-year from the state in order to facilitate these participatory learning experiments, as well as ongoing pressure and vigilance from civil society. The forum, then, is a specific organized encounter that may be a one-off or part of a series of events, and which emerges from and reshapes learning assemblages.

When we switch to the translocal context of urban forums across a global North–South divide, distinct issues arise. As the SPARC/Groundswell temporary forum revealed, the categorizations of North–South, with their attendant stereotypical connotations, can militate against the prospects of learning through difference. In contrast, other examples from the literature have shown how exchanges can serve to disrupt such geographical categorizations and thereby encourage learning (e.g. Gaventa 1999). In this context, I have argued that the broad conception of learning through translation can be useful if the possibilities of translocal learning are to be enhanced rather than restricted to a reductive conception of learning as direct transfer between 'similar' cities. If the stumbling block in the example highlighted is the stubborn persistence of the global North–South divide, the more general issue at stake here is the importance of learning through difference over attempts to learn through pre-given criteria of similarity. The next chapter develops this dialectic of similarity and difference through a critical discussion of the translation of urban policy and planning initiatives across space-time.

Chapter Five

Travelling Policies, Ideological Assemblages

Introduction

Recent years have seen an upsurge of interest in travelling forms of urban policy and planning. In particular, an emerging 'policy mobilities' approach has sought to critically examine the translation of urban policy as it increasingly travels translocally. In this chapter, in concert with much of this literature (e.g. Peck and Theodore 2010), I argue that ideology plays a crucial role both in structuring the sorts of policy ideas that become mobile, and in shaping the forms of learning that take place through policy relationalities. Building on this, I aim to contribute to this literature in two ways. Firstly, I offer a critical framework for conceptualizing urban policy mobilities that highlights four key issues: first, the *power* at work in policy learning, i.e. the forms of power that promote, frame or structure particular kinds of learning; second, the *object* of learning, i.e. the epistemic problem-spaces that the mobility of policy and planning creates and addresses; third, the *form* of learning, i.e. the organizational nature of learning; and, fourth, the *imaginary* at work in learning, i.e. the image of urban reassembling that learning seeks to accomplish. The aim of this framework is to critically examine the ideologies and inequities of mobile policy learning, as well as the role of the specific agents that constitute these urban learning assemblages. While I develop this framework in the context of policy and planning debates, it is applicable more widely to forms of urban learning, for example in relation to social movements like SDI (Chapter 3) or within urban learning forums (Chapter 4). I use it in this chapter to critically expose the ideologies that structure learning assemblages in their translation, coordination and dwelling.

Learning the City: Knowledge and Translocal Assemblage, First Edition. Colin McFarlane.
© 2011 Colin McFarlane. Published 2011 by Blackwell Publishing Ltd.

Secondly, in developing this framework, I aim to unsettle the presentism of urban policy mobility scholarship. While scholars working in this area would no doubt acknowledge the historical role of translocal policy learning in shaping cities, the tendency in this literature is to focus on contemporary policy mobilities. If the extent and speed of policy mobility has increased – driven by the growing amount of travel, conferences, study tours, policy networks, and the Internet – urban policy mobility is, of course, far from new. Planners, architects, policy-makers and consultants have *always* sought to learn from elsewhere in their attempts to assemble the city, so much so that the city is always already a relational product of different agendas and strategies from other cities. In order to demonstrate different contexts, logics, and forms of mobile urbanism, I examine quite specific urban learning assemblages from three particular periods. Firstly, I discuss learning as translation in the example of colonial urban learning. Focusing on urban planning in colonial Bombay (now Mumbai), I argue that comparison – as a central form of learning as translation (Chapter 1) – was central to how urban planners learnt the city. This urban learning assemblage was predicated upon inhabiting an idea of the 'decontaminated' city, a profoundly material conception of the city as cleaning polluted urban natures through infrastructure technologies. This example also reveals how urban planners learnt *the city as such*, i.e. at stake in urban comparative learning was not just forms of planning, but the existence of Bombay *as a city*.

Secondly, I discuss mobile urban planning initiatives in the 1950s driven by Cold War ideologies of modernism and socialist realism, taking the example of the remarkably influential architect Constantinos Doxiadis and his plans for Baghdad, as well as deeply ideological planning initiatives for East Berlin during the German Democratic Republic (GDR). Doxiadis and his visual plans became a central actor in a translocal learning assemblage that combined scientific modernist planning with a capitalist vision of the city. In contrast, socialist realism explicitly opposed modernist planning, and in East Berlin sought to propagate a socialist vision of the city that combined local architectural traditions with Soviet monumentalism. Again, at stake here in both the examples of scientific modernism and socialist realism is not just a particular urban plan, but a conception of the city as such. As with colonial Bombay, these are sociomaterial conceptions of the city that sought to reassemble urbanism through an ideologically structured form of learning.

Thirdly, I consider the nature and implications of neoliberal ideology both for the kinds of urban policy that become mobile, and for ways in which learning is framed, translated and contested. I draw on a variety of examples here that illustrate how ideologies of urban 'development' (e.g. Harare and Guayaquil), urban 'recovery' (e.g. post-Katrina New Orleans), and of the 'creative' or 'smart' city, structure contemporary urban policy learning assemblages. Particular agents become important in framing these neoliberal forms of mobile urban policy, including institutions like the World Bank or the Manhattan Institute, or influential individual consultants like Richard Florida and William Bratton. In the three different periods the chapter examines, my selection of cases is, to be sure, selective. But the choice is not random: the examples chosen powerfully reveal how ideology can shape the nature of urban learning, and the very range of urbanisms examined in this chapter is an attempt to show that the critical framework of reassembling policy urbanism that the chapter deploys – attuned to power, object, form and imaginary – is one that can be usefully deployed to grasp how learning from elsewhere plays an important role in how cities are reassembled over time. The three contrasting periods also serve to remind us that if policy mobility is increasing in its extent and speed, we need to be careful not to assume that it therefore possesses a greater intensity. The effect of policy mobility is contingent, and it does not need to be fast or ready-to-hand to have a transformatory impact. For example, the relatively slow movement of ideas or infrastructure materials from Britain to India in the mid-nineteenth century did not mean the influence of the ideas was less intense – there is no necessary pre-given relationship between the nature and speed of policy movement and its resultant effects. The chapter concludes with a reflection on the limits of focusing on ideology as a framing device for urban learning.

Translating Policy

There is a long history in urban studies of tracing policy mobilities, whether in the shape of urban policy transfer, such as Anthony Sutcliffe's (1981) *Towards the Planned City*, or Ian Masser and Richard Williams' collection (1986) *Learning from Other Countries: The Cross-National Dimension in Urban Policy-Making*, to Anthony King's (1976, 1991a,b) surveys of colonial urbanism, Joe Nasr and Mercedes Volait's (2003) collection *Urbanism: Imported or Exported?*, Healey and Upton's (2010) collection *Crossing Borders: International Exchange and Planning Practices*, or McCann and Ward's (2011) collection, *Mobile Urbanism*. In conventional policy transfer literature, there has been a tendency to conceptualize learning as a rational transfer of knowledge and ideas, detached from relations of power that effect mobility and translation (Rose 1991, 1993; Dolowitz and Marsh 1996; Stead *et al.* 2010). Richard Rose (1991, 1993), a prominent voice in policy transfer approaches, portrayed policy travel as a rational process of selecting best-fit policies off the shelf: identify a goal, peruse the available policy choices, and implement the policy. In recent years, however, a new body of work has emerged critically examining the increasing travelling of urban policy (e.g. Ward 2006, 2007; McNeill 2009; McCann 2008, 2011; Bunnell and Das 2010; Peck and Theodore 2010; Larner and Laurie 2010;

McCann and Ward 2010, 2011). If conventional policy transfer literature often assumes a linear transfer of unchanged knowledge from one setting to another, the emerging literature on what Eugene McCann (2008, forthcoming) calls 'urban policy mobilities' is concerned with how policies are framed and translated through ideological alignments and power relations, and with how knowledge is transformed through the messy realities of local histories and concerns, embodied practices and materialities. This reveals a non-linear geography of policy travel through diverse origins and routes, including blogs, conferences, consultants, chats over coffee, glossy documents, PowerPoint™ presentations, policy fads, historical networks, and the ideological framing of problems and solutions – in short, the 'connective tissue' that constitutes urban policy mobilities (McCann 2011: 109; Friedmann 2010). Here, apparently banal activities ranging from phone calls and e-mails to bus journeys and lunches play sometimes important roles in constituting the relations between people and sites and in shaping how learning emerges and changes (McCann forthcoming). Such an approach is suited to assemblage thinking through its focus not just on relational composition, but more specifically on *incorporated externalities*, i.e. on how a mixture of space-times are assembled into a particular way of seeing an urban problem and solution.

While the emerging policy mobility literature offers an important step forward here through attempting to examine the ideologies, power relations and messy realities that constitute travelling policies, an important contribution of Rose's work was to underline how historical relationships can dictate the sorts of 'lesson drawing' that policy-makers embark upon. In the case of the UK, for instance, he argued that policy-makers appear almost instinctively to look to countries such as Canada, Australia, New Zealand and particularly the USA when attempting to draw lessons about policy. These are countries where, as Rose argued, policy-makers perceive 'common values' – for instance of capitalist democracy or culture – to be constant: 'Elected officials searching for lessons prefer to turn to those whose overall political values are consistent with their own' (Rose 1991: 17). Dolowitz and Marsh (1996: 353), while similarly underestimating unequal power relations and conceiving policy transfer in a similarly rationalist sense that imagines knowledge travelling from place to place instrumentally, echoed Rose's argument: 'If policy makers are looking to draw lessons from politics which are similar in institutional, economic and cultural makeup, it might be argued that, instead of expanding the number of ideas and actors involved in the decision making process, policy transfer enhances the power of a relatively small circle of actors who consistently draw lessons from each other.' This critical perspective on learning from the usual suspects is an important contribution to take from the policy transfer literature, and one that I will return to repeatedly in the examples in this chapter. It is not an insight that has newly emerged through critical urban policy mobility approaches, but which has been taken up by them. As Peck and Theodore (2010: 171) argue, if translocal policy learning ostensibly appears as a pragmatic form of exchange, in practice it often operates in the context of narrow ideological parameters, within which there is a concerted technocratic framing toward certain favoured solutions: 'The reason why many models achieve mobility in the first place is that they have, in one way or another, been ideologically anointed or sanctioned.'

The critical urban policy mobility literature echoes a wider shift in urban studies in how urbanism is conceptualized, revealing the city – as Ash Amin (2002: 397) has put it – as 'a place of engagement in plural politics and multiple spatialities of involvement'. The recent urban policy mobility turn, then, is part of a much broader effort to conceive the city as a relational product, work that in recent years has examined not just urban policy mobilities, but – to name just a few areas – 'boomerang' military urbanisms (e.g. Graham 2009), travelling forms of urban revanchism and punitive urbanisms (MacLeod 2002; Baviskar 2002; Peck 2003; Swanson 2007; Wacquant 2008), and the circulation of neoliberal ideology more generally (e.g. Peck and Theodore 2001, 2008; Peck 2005, 2006). Despite their substantive differences, there is a shared relational perspective in these accounts that argues that it is impossible to understand cities as territories *prior* to their engagements with other places (Amin and Thrift 2002; Massey forthcoming, 2007). A key challenge here is to critically examine the forms of power at work in shaping translocal learning assemblages.

Given that the routes through which urban policy travel are often those of élite epistemic communities, there is inherent exclusivity to policy learning. Peter Haas (1992) described epistemic communities as collectives of experts in a particular policy domain that share: normative and principled beliefs which provide value-based rationales for their action; common causal beliefs or professional judgements; common notions of validity based on inter-subjective, internally-defined criteria for validating knowledge; and a common policy enterprise. These formations echo work on 'transgovernmental networks' (Risse-Kappen 1995: 4), 'networks among state officials in sub-units of national governments, [and] international organizations'. Both epistemic communities and transgovernmental networks have a major impact on the translocal translation of values, norms and ideas, but these approaches often stop short of tracing the contingencies, non-linear processes, ideologies and multiple forms of power through which knowledge travels and learning occurs, and which are an important focus of emerging urban policy mobility debates.

John Allen's (2003, 2004) work is useful for identifying a range of different powers of specific character and reach that shape how knowledge is translated through multiple space-times. While avoiding a narrow spatial determinism, Allen argued that spatiality is constitutive of the particular ways in which different modes of power take effect. Here, the challenge is to attend to the material contexts of particular places and to the range of spatialities that connect 'here' and 'there', because they are of substantive importance to how power takes shape. Allen's (2003, 2004) work usefully points to a range of different powers, including domination, authority, manipulation and seduction, all of which are different in their character and reach. Domination works quickly to close down choices and may be more effective across distance, while authority works most effectively through proximity and presence, drawing people into line on a daily basis and seeking the internalization of particular norms. Authority's need for constant recognition means that the more direct the presence, the more direct the impact. Conversely, the larger the number of outside interests to negotiate, the more varied the mix of resources, the greater the potential for authority to be disrupted. Manipulation can have a greater spatial reach than authority partly because it may involve the concealment of intent, such as in an advertising campaign or corporate development intervention, and partly because it does not require the internalization of norms. Seduction is a more modest form of power that can operate successfully with spatial reach, 'where the possibility of rejection or indifference are central to its exercise' (Allen 2004: 25). These different modes of power are mediated in space and time, so that manipulation may become, for instance, seduction or authority.

There are of course many other modes of power. Inducement may involve financial incentives to obtain compliance, such as in state contracting of urban NGOs (Ferguson and Gupta 2002; McFarlane 2008c, 2009a), while coercion may include monitoring or target-setting, for example in donor aid monitoring of states or NGOs (Mawdsley *et al.* 2001). Power may be instrumental, a series of actions designed to make others act in ways that they would otherwise have not, or associational, involving the formation of a common will. Seen in the context of Allen's topological conception of the geographies of power, the relations between power, space and learning multiply. Learning may involve close and repeated authoritative monitoring, or the more distant closing down of alternative knowledges and ways of seeing through domination. It may involve inducement through rewards based on learning, or it may be inspired by translocal solidarities in the formation of common will. These different modes of power work alongside, build on, and extend more commonplace accounts of power over distance as hegemony (the ideological fermenting of consensus through coercion and consent) or governmentality (the delineation of 'problems' and 'solutions' that are to be met through particular techniques that regulate conduct), but most importantly they emphasize the multiple and often simultaneous transformation of power across space, and it is in this sense that it is useful for a relational topology of translocal urban learning assemblages, that is, how they are translated, coordinated and dwelt. I will deploy

Allen's typology of power throughout the chapter in examining how policy and planning learning occurs through mobility.

The urban policy mobility literature represents an important opportunity to open up the 'black box' of urban neoliberalization and to uncover how policy learning is constituted in practice (Ward 2006). But while the authors driving this work are, of course, aware of the importance of history in shaping the trajectory of contemporary urban policy assemblages, to date this work has had a tendency to focus on the contemporary. The unintended consequence is that the reader can be left with a sense that urban policy mobilities are somehow new. In examining particular episodes of translocal urban policy learning from the mid-nineteenth century, the early post-Second World War years, and the contemporary period, this chapter seeks both to unsettle this presentist emphasis, and to consider how learning occurs in these different contexts (see also Clarke forthcoming). We live in a period of 'fast policy transfer' (Peck and Theodore 2001, 2010), but there are long and varied histories which, notwithstanding their different nature and contexts, can be understood through a critical framework of urban policy learning.

As Pierre-Yves Saunier (e.g. 1999, 2001, 2002) reminds us, far from being 'new', urban policy has long been on the move. The emergence of a municipal 'urban internationale' – formed in urban France and initially linking European and North American cities but by the 1940s also connecting South American, Asian and African cities – in the early twentieth century is testament to this, driven by the philanthropic spirit of leading urbanists like Patrick Geddes, Paul Otlet and Albert Thomas, and supported by institutions like the Carnegie Trust, the Rockefeller Foundation, the Ford Foundation, the International Labour Office, and the League of Nations. For Saunier (2001) it was not technological advancements like transatlantic telephones, cheaper air travel or the growth of cities that drove these translocal exchanges, but a sense amongst these individuals and institutions that urbanism constituted both the urban future and the locus of progressive change. In doing so, this disparate movement framed the city – as a global phenomenon – as in need of a universalist and expert-driven modernization, and connected in particular to debates around the modernization of planning, housing and government in the United States. As Saunier (2002: 520) has written, the underlying faith driving these connections was: 'If towns are the planet's future, if contemporary urbanization faces political and technical authorities in every city with the same problems, then the exchange of experience becomes a meaningful activity and its results can transcend frontiers. This was the message of the first congress of the Union *Internationale des Villes* in 1913 – and of the Transatlantic Summit of mayors in 2000' (and see Nick Clarke's work on municipal twinning and exchanges, 2009, 2010).

Comparative Learning: Translation and Colonial Urbanism

As Nasr and Volait (2003) showed, there is nothing new about translocal urban policy learning. The literature on colonial urban planning is instructive here. Bombay, for example, like other Indian port cities such as Calcutta and Madras, and like many other cities across European empires, was always conceived in part through comparative learning, as a hybrid city developed through European discourses of planning and improvement (Headrick 1988; Rabinow 1989; Wright 1991; Chattopadhyay 2009; King 2004; Hosagrahar 2006; Legg 2007, 2008; Harris 2008; McFarlane 2008a). Writing about colonial Lagos, Gandy (2006: 375) described how British colonial administrators 'sought to transform the port into the "Liverpool of West Africa"'. Perera (2005) narrated how in colonial Colombo, modified British town planning discourse, mediated through a variety of agents, including influential individuals like Patrick Geddes and legislation like the 1915 Housing Ordinance, laid a particular view of the capitalist city over the colonial city, marginalizing the poor. As he argued, British experts 'saw what they knew', problems familiar to British industrial cities, and proposed plans that were effectively futures for urban Britain, rather than urban Sri Lanka. Learning was central to the colonial project more generally as an encounter. For example, Derek Gregory (2000) described attempts by colonial travellers to turn a scene of apparent Oriental confusion – the street bazaars of Bombay or Cairo, for instance – into a knowable regularity or hierarchy. Gregory (2000: 308, 313) described 'the viewing [often masculinist] subject's active desire to know, to possess the Orient as an object of knowledge' – often through a process of 'learning-by-rote, a proto-Taylorist discipline of cities, sites and sights' – at the centre of the Orientalist epistemology of British Victorian and Edwardian high culture.

Urban planners were not immune to this desire to possess-through-learning the colonial cities they administered. Indeed, as Harris (2008) argues, the labels that postcolonial scholars often use today – métissage and creolization in the Caribbean, transculturation in Latin America, and hybridization in India – shot through with deeply unequal relations of power, were developed in order to grasp forms of relationality practised by planners and others in colonial contexts (Young 2001). These attempts to grapple with colonial mixture have witnessed important discussions on the nature of power and change as different ideas, objects and people interact. In this section, I draw upon the example of one instance of colonial comparative learning in the context of urban planning in mid-nineteenth century Bombay: the planning of urban sanitation in what was viewed as a city contaminated with human and animal waste. As Chapter 1 argued, urban comparison involves the translation of different urban examples, and is central to translocal learning.

In examining the travelling of colonial urban planning, I want to highlight one influential report produced by a key public health official in mid-nineteenth-century Bombay, Henry Conybeare. Conybeare was the Superintendent of Repairs to the Board of Conservancy in Bombay in the 1850s.[1] His 1852 report to the Board, *Report on the Sanitary State and Sanitary Requirements of Bombay*, was an important contribution to municipal government understanding of sanitation (Dossal 1991). I have chosen to highlight this report for two reasons. Firstly, it was influential in conceiving sanitation as a problem and public health as a set of solutions in urban government circles. Sanitary reform was championed by major local officials such as Conybeare, as well as other influential figures of the time such as Arthur Crawford and Thomas G. Hewlett (both Municipal Commissioners), and Andrew Leith, Inspector General of Hospitals, who were 'deeply influenced by the public health movement gaining ground in Victorian Britain' (Dossal 1991: 125; McFarlane 2008c). Conybeare believed in the duty of municipal bodies and governments to provide healthy living environments, and identified himself with British reformers like Edwin Chadwick and John Simon. Secondly, the report makes its arguments through drawing explicitly on colonial comparison as a central means for understanding sanitation as problem and public health as solution. Sanitation solutions, such as drainage, were conceived relationally, and in this learning assemblage a clear hierarchy is maintained between the metropole and the colony. Many of his recommendations – and those of others in government – were at odds with popular sanitary practice in the city (and see Hosagrahar 2006 on Delhi). The consequence is that comparative learning features as an important metric for governance, and frames policy discourse around the possibilities of resource, expertise and institutions.

In responding to what he saw as a contaminated city, British technology, especially sewers, latrines and water inflows, were of critical concern for Conybeare. But it was drainage specifically, in his view, where investment had to be focused and health improvements and mortality reduction could be most readily achieved. Conybeare acknowledged, however, that he lacked data from Bombay to support his case on the importance of drainage. The argument he subsequently made rested upon remarkably little reference to Bombay itself. His emphasis was less on Bombay's geography to demonstrate the need for drain improvement, and more on the metropole. It is ironic that Conybeare did so in part because of the lack of knowledge about Bombay, thereby perpetuating the problem:

> We are, indeed, without the first basis for sanitary statistics – a trustworthy census, an annual mortality return; our town, with half a million of inhabitants,

---

[1] The Board of Conservancy was set up in 1845 to coordinate the growing administration of the city. Its remit involved regulating civil and material infrastructure, with the Government of Bombay Presidency retaining final decision-making capacities. The Board of Conservancy laid the basis of municipal organization in the city and was a forerunner of the Bombay Municipal Corporation, set up in 1873 (Dossal 1991).

is not divided (as for half a dozen municipal and sanitary purposes it ought to be) into any generally recognized districts and sub-divisions; the houses have one number for police and census, and another number for house assessment. There is, in fact, a general want of unity and system. (1852: 2)

The Native Town was particularly poorly known. Of the Native Town's neighbourhoods, the elite 'generally know as little as they do of the interior of Africa' (ibid.). The lack of local data meant that Conybeare relied on English sanitary statistics 'to show what would be the effect of sanitary improvements in diminishing the annual death-rate of Bombay' (1852: 3). While Dossal (1991: 128) argued that Conybeare's report provided a 'vivid description of the hopelessly inadequate drainage and sewage system which existed in Bombay town in the early 1850s', the emphasis on colonial comparison means that there is for large parts of the report (142 pages in length) strikingly little about the city itself.

For Conybeare, the differences between England and Bombay did not diminish at all the utility of the comparison – Bombay was more overcrowded than an English town, and if open drains and cesspools have an impact on health and mortality in England, then it can only be more the case in a tropical climate with a monsoon. Conybeare argued that if in London a geography of drainage maps on to a geography of mortality, fever and illness, so it must apply to Bombay, and here he returns to Bombay to point to differences between the élite A division (Fort, Esplanade and Colaba) to the south compared with E division (Mazagon, Tarway, Cammatee Poora, Parell and Sewree) to the north as differences between drained and undrained spaces. Conybeare was using British sanitary precepts as a basis for learning – and in the process actively producing – Bombay's sanitary geography. Sanitary statistics from English towns demonstrated that 'a large and specific amount' of death is from the 'deficiency of covered drains' – about 20 per cent of excess deaths are attributable to this alone, Conybeare (1852: 17) argued. Long extracts on Manchester and Charlton are presented as speaking for themselves, as if resolving the issue alone. While he lived in Bombay, he chose not to embark on local research, but to learn the problems and solutions of sanitation in Bombay by inhabiting a way of seeing based on knowledge from elsewhere.

Conybeare and other colonial officials believed that 'natives' had to have sanitation enforced upon them, and doubted that they would voluntarily follow norms of public health, i.e. the Chadwickian emphasis on locales rather than people as the locus of disease, and on the need for drainage and personal hygiene. Drawing upon police evidence from the UK, Conybeare (1952: 22) argued that 'sanitary reform is in itself a police improvement; and that crime, dirt, and a high-rate of mortality, are generally found to be co-extensive'. Here, Conybeare drew direct links between 'filth' and 'immorality', highlighting parts of Whitechapel and Glasgow, where disease and crime inhabit similar spaces.[2] This association between police and sanitation was probably part of an attempt by Conybeare to raise the profile of sanitation on the municipal planning agenda, but it was more than this. It also reflected a colonial ideology of cultural supremacy. While in the UK there was a certain degree of faith that people would adopt more sanitary behaviour as public consciousness grew – including ensuring clean and dirty waters were kept separate, following regulations on the provision of latrines in new developments, and not defecating in public – in Bombay, Conybeare, among others, had no such faith in what fellow municipal official Andrew Leith (1864: 25) referred to as a population 'slow to believe'. If Conybeare sought a reduction in police spending in favour of public health, he none the less wanted sanitation improvements to be policed through municipal enforcement.

In raising this issue of comparison, I am less concerned with questions of conceptual robustness than with what this move indicates about the discursive power of comparison as a mode of learning in urban colonial government. Conybeare's report is littered with comparisons as often the most important basis for understanding and gauging the extent of a 'problem' and the justification of what needs done and how it should be done. In the report at least, he does not question the logic of understanding the contaminated city through living conditions in urban Britain and the gathering momentum of the British sanitation movement, rather than through Bombay and its inhabitants. Bombay was often grasped in the shadows of British planning discourses. The mode of colonial comparison deployed by Conybeare, one that understands, measures and seeks to make a place reconcilable with a British model, underlines the legitimacy of the imperial interest while marginalizing the views and practices of the city's inhabitants (as opposed to forms of hybridity that may challenge colonial power; Morton 2000).

At stake in this example of colonial urban learning was not just a particular policy intervention, but a conception of the city as such. There are two ways in which comparative learning mapped onto this broader production of the city. Firstly, urban comparison was a basis for producing Bombay as a *modern* city. As Chakrabarty (2002: 66) argued, in bringing together concerns with order, police, civic consciousness and a particular kind of aesthetic, debates about public health and hygiene cannot be separated from the broader project of modernity. Reactions to the contaminated city as a place characterized simultaneously by 'cesspools', miasma and disease, as well as narrow streets and perceived criminal or rebellious behaviour, point to a particular way of seeing urban public space

---

[2] His view on morality and sanitation resonates with the impetus of a great deal of reform across not just the British but other European empires (see Wright 1991, on the discursive linking of sanitation and overcrowding with morality and aesthetic squalor in Paris and the French colonies).

premised on a certain notion of the modern city. Conybeare, amongst others, recommended that the 'labyrinth of crooked narrow alleys in the Indian quarter be cleared and straight streets run through it, which would facilitate the laying of drains and faster flow of traffic, and improve ventilation' (Dossal 1991: 130–1). In this view, public health involves an enrolment of nature, the social and infrastructure, with infrastructure playing a crucial role in a broader context of conceiving and measuring improvement. As Sharan (2006: 4906) has suggested in reference to colonial Delhi: 'Infrastructure in the colonial city, it may be suggested, operated most powerfully in the symbolic realm, gesturing to an imminent modernity, even as that modernity was endlessly deferred.' If modernity was deferred, modernist categories of public and private were also subverted by the ways in which Indians used open space, from washing to changing, to sleeping, and urinating and defecating in the open. This tested the patience of reformers who sought to cajole new subjectivities through networks (see Otter 2004 on London). Andrew Leith (1864: 15), for instance, complained of 'indecent' habits in Bombay's public spaces that the government found difficult to end: 'There is scarcely a part of the Fort or Native Town in which the ground along every dead wall is not wet or in pools from its being resorted to as an urinary ... regardless of decency, and this custom is unchecked.' This is despite boards bearing the threat of penalties, 'as if in contempt of these mere declarations of what the law is, nuisances of the most odious kind are daily or nightly committed under them' (Leith 1864: 16). These spaces were opposed to the order of the European quarter, especially in the areas of the Fort, Colaba and Malabar Hill, containing large, gothic buildings, wide streets and open parks (London 2002). As Chakrabarty (2002: 68) argued, opposing these spaces comparatively reflects the perennial gap between modernist desire and popular practice. Comparative learning was crucial to how administrators came to see the city as a modern city.

Secondly, urban comparison entailed not just learning about sanitation infrastructures for Bombay, but learning how to govern the city. There are two steps in this: firstly, the 'recognition' through comparison that Bombay does not function *as a city* because of its differences from British cities. Conybeare compared Bombay with British cities in order to calibrate the progress made in those respective sites with sanitation. In Britain, he argued, the sanitation movement had progressed through three domains: first, in *legislation*; second, through the establishment of *local government* in municipal establishments in most large towns; and third, in the reduction of *costs* of drainage through the development of sanitary engineering in the work of Parliamentary Commissions. Conybeare compared each of these domains in England with the situation in Bombay, and complained that Bombay did not meet the standards set in England. For example, the enforcement of local building Acts in England was more stringent, creating 'a greater regularity of streets and buildings in English towns' (1852: 8). Conybeare examined the 'applicability' of English improvements to India and Bombay specifically, using as standards 'the improved system of municipal establishments, sanitary enactments, and sanitary engineering, now generally adopted in large English towns', including Plymouth, Liverpool and London, and found Bombay wanting in each area (1852: 10). He complained that Bombay had only a fifth of the average proportion of sewerage to population in England. He went on (1852: 26): 'In Bombay, on the other hand, the sewers are not water-tight, and are all laid close to the surface, in soil which, owing to the dryness of the climate, sucks up every drop of moisture within reach of it', all of which was compounded by an absence of house-drainage. Improved house-drainage would allow waste water to be added to the sewers, increasing flow, instead of water being thrown out and drying up in the soils, and thus create sewers more in tune with how they function in Britain.

The second element in the development of a conception of the city as a governable space is the production of government itself. The contaminated city as a problem, and public health as a solution, was not just the *domain* of government, but was *productive* of government. Officials sought to learn about the city through data collection techniques such as the census or mortuary returns, and through conceptual lenses for creating and organizing knowledge, visioning the city and conceiving improvement. The broad rubric of public health was one important organizational frame here, and sanitation infrastructure was at the core of this. Other frames emerged around town planning (see Perera 2005 on Colombo), congestion (see Legg 2007 on Delhi), law and order, military security, transport, and gas. That these different frames were coordinated was important to how priorities were drawn up, and to the shift in perception of Bombay as a trading port to a city in its own right.

How does this example of comparative learning relate to the framework for assessing urban policy learning mentioned at the start of the chapter, i.e. the focus on power-object-form-imaginary? The form of *power* at work in this case of colonial comparative learning, following Allen's (2004) account, was a combination of seduction (of the technological promise of the modern, clean city), inducement (for example, fines to try to ensure compliance, even if this form of power was often resisted or ignored), and instrumental power (a series of infrastructures designed to make others act in a certain way, such as using latrines, or disposing of waste water in drains). At times, these forms of power would co-exist or translate into one another. For example, inducement and instrumental power may become authoritative power in the form of police enforcements. The *object* of learning was to instil a conception of the clean, modern, Westernized, circulatory city, while the *organizational form* of learning was that of hierarchical comparativism through reading British urban sanitation and public health documentation.

The *imaginary* at work here is that of the contaminated city made clean, ordered, modern, and more familiar to British eyes. Taken together, the power, object, form and imaginary of comparative urban learning in this case was that of a hierarchical ideology of Western colonial modernism, where British knowledge and technologies rather than everyday Indian urban dwelling constitutes the reference point for learning.

Ideology and Postwar Urban Planning

In the years following the Second World War, variants of ideological colonial urbanism were pursued through a Cold War urban geopolitics that shaped urban learning assemblages of policy-makers, architects and planners. In this section, I focus on two important examples of this: the work of the infamous Greek planner and architect Constantinos Doxiadis, and the socialist planning of East Berlin. In the 1950s and 1960s, Doxiadis was an important figure in US Cold War urban policy, designing new towns throughout the Middle East and Africa which aimed at inculcating 'democratic' and 'free-market' values, and which largely ignored indigenous traditions. Doxiadis pursued what he called 'Ekistics' (the name of a journal he founded in 1955), the science of human settlements: gridiron modernist plans designed for cities in places as different as Ghana, Zambia, Sudan, Pakistan, Iraq and the US, all enabled by the support he received from the Ford Foundation (Provoost 2006). Echoing the colonial desire to possess-through-knowing, the charts and statistics he provided, informed by biological metaphors of arteries, hearts and flow to provide a network conception of urbanism, created the illusion of control through planning (Wigley 2001). Indeed, the network idea was at the heart of how Doxiadis sought to learn and propagate his form of planning and architecture. For example, he used a spider's web as a model of the ideal network city, and in 1963 he brought together for the first of several meetings a range of leading academics and practitioners, including Marshall McLuhan and Buckminster Fuller (later President of the World Society for Ekistics), onboard a ship off the Greek coast – a networking operation that created Ekistics missionaries of the participants (Wigley 2001).

Doxiadis, argued Mark Wigley (2001), was obsessed by global urban spread. A crucial set of networks for Doxiadis emerged through his work with the Ford Foundation, which enabled a marriage between the Foundation's passionate anti-communism and Doxiadis' scientific modernism. The links with the Ford Foundation, which began with work on Karachi in the mid-1950s, emerged partly from what Provoost (2006) called Doxiadis' considerable charisma and ability to work political networks – what, following Hansen and Verkaaik (2009), we could call his urban infrapower (Chapter 2). The Ford Foundation provided an assemblage that incorporated strands of US foreign policy, businesses and universities, including the joint Harvard-MIT Centre for Urban Studies that was sponsored by the Ford Foundation. This created a constant feedback process, an urban learning assemblage not restricted to any particular city but unified through a powerful ideology of urbanism as capitalist democratic 'freedom' and scientific modernism. One example was Doxiadis' role in planning Baghdad, enabled both by the Ford Foundation and an Iraqi regime eager to embrace Western modernization, but to do so away from its longstanding British connections.

A key element in the constitution of this mobile urbanism was Doxiadis' powerful new images, which offered potent visual inscriptions of his brand of urban planning as a science, and of the future city: 'He presented the outcome of his studies and designs in grids, charts, diagrams, and schemes', appearing apparently objective and rational (Provoost 2006). When combined with a seemingly scientific 'formula', the *imaginary* at work in Doxiadis' plans was to prove more seductive than the plans for Baghdad that rival American or European firms were advocating, as Theodosis (2008: 169) writes: 'The managerial skills of Doxiadis and the multidisciplinary approach of Ekistics, a planning methodology that Doxiadis coined the science of human settlements, composed an appealing formula that the Iraqi state couldn't bypass.' The *power* at work in learning here, in closing down the possibility of alternative urbanism, took the form of an authoritative discourse of planning-as-science. Ekistics was presented not as a model to be discussed through dialogic learning with local groups and indigenous urbanisms, but as the already learned 'truth' of urban development. This scientific modernist approach, of course, connected strongly with the Ford Foundation's goals of educating non-Western people into 'rational' and therefore 'civilized' peoples, and of opposing Soviet styles. For the Ford Foundation, this goal of conversion constituted the *object* of urban learning in Doxiadis' extraordinarily pervasive 'vehicular urbanism' (McLennan 2004; Peck forthcoming). As Chapter 3 demonstrated in relation to social movements, materialities such as documents, images and models can play a crucial role in facilitating the influence of certain forms of urban learning. The *organizational form* that propelled this work took the shape of conferences, journals, models, plans, charts and diagrams.

One Doxiadis-designed site is now a focal point in a contemporary war of a different sort: Sadr City, Baghdad's sprawling slum settlement on the east bank of the Tigris. Sadr City – renamed from Madinat al-Thawra, the 'City of the Revolution' – was planned as a large repetition of square, concrete neighbourhood units; low-rise, high-density developments with narrow alleyways. It was built from 1958 as part of Doxiadis' masterplan for Baghdad, set around community centres of market buildings, public services and mosques. Some local techniques and building methods were integrated into the plans, including panels decorated in broadly cast Arab motifs, but local cultural urban traditions and identities were largely ignored. Doxiadis' urbanism, which he would later call *Dynopolis* ('dynamic city'), failed to recognize the public role of the old city's urban density and overlooked the fact that the streets of the old city, despite their narrowness and darkness, had immense social value (Pyla 2008). In addition, the plans 'remained bounded by a rather linear hierarchical logic that assumed that communities and subcommunities could neatly fit into each other' (Pyla 2008: 12). For instance, Pyla (2008: 15) argued that the attention to so-called 'gossip squares', a key element of 'local' tradition in the plans, were misguided because they 'catered more to an orientalist nostalgia than any profound understanding of Iraq's public life, the intense heterogeneity of its society, or its aspirations to modernity. Still overpowered by the modular functional plan, the gossip squares, hamams, and mosques were subsumed by the grand formal and social order of the master plan.' In addition, the microclimate of housing units compared unfavourably with the old city's mud huts with their movable roofs. Tokenistic as it was, the attempt to incorporate local traditions none the less reflected at least *some* awareness that the 'success' of urban planning for capitalist democracy rested to some extent not on the authoritative voice that had originally so impressed Iraqi administrators, but on the *associational* formation of common will through learning by translation.

Indeed, the attempt to promote a particular brand of urban learning through the authority of planning-as-science was translated in ways that Doxiadis and the Ford Foundation may not have predicted. As Theodosis (2008) suggested, the delay between Doxiadis' master plan and the construction of housing in Sadr City resulted in significant differences in housing style and density as local actors tinkered with plans on the ground. More generally, the plans failed to address community cohesion because they started not with how different people use urban public spaces in Baghdad, but with debates on constructing public space in Western planning. The plans also failed to take account of future growth of the city, which left new immigrants with little alternative to informal housing on the periphery of the city. Moreover, the renewed debates around architecture and planning gave rise to an architectural movement in the 1960s in Iraq that sought a renewed focus on local building traditions. The space-times that constitute learning assembles are, while ideologically framed, multiple and non-linear in nature.

While Doxiadis was promoting his brand of urban reassembly on behalf of one ideological half of the Cold War, socialist realism, promoted by Stalin and explicitly opposed to modernist planning, was becoming dominant throughout the Soviet Union, particularly up until Stalin's death in 1953 (Strobel 2003). In East Berlin, socialist realism translated as highrise constructions – or more commonly 'mid-rise', as skyscrapers were viewed as antisocial in scale – without traditional Berlin courtyards, which were viewed as unhygienic and too dense (Strobel 2003: 133). This was combined with a tendency both to monumentalism and classicism permitting decorative detail. An important example here is the grand yet intricate Stalin Allee (now Karl Marx Allee, see Figure 7), East Berlin's most spectacular street at 90 metres wide and two kilometres long. Karl Marx Allee, once described by Italian architect Aldo Rossi as 'Europe's last great street', was built by the young GDR between 1952 and 1960, running through Friedrichshain to Mitte, and was a flagship building project of East Germany's reconstruction programme after the Second World War. The length and breadth of the avenue was not just for aesthetic effect, but for use by tanks and for East Germany's annual May Day parade. It had an extraordinarily important symbolism and was part of a wider plan to develop a new socialist centre in East Berlin along the lines of the 1935 General Plan for Moscow of monumentalizing the city centre, most of which was never realised. The avenue was designed by several architects – each responsible for different parts of the sections – with the aim of blending the monumentalism that characterized much early postwar Socialist architecture with Berlin's specific architectural traditions. It is lined with eight-storey buildings designed in the *zuckerbackerstil* or 'wedding-cake' style of socialist realism seen in several parts of the Soviet Union. The attempt to architecturally translate the buildings through Berlin's history, however, means that they differ from those in Moscow. In particular, the facades often contain traditional Berlin motifs. Most are covered by architectural ceramics and contain pillars, a nod to the city's architectural history and especially to the neoclassism of Karl Friedrich Schinkel, the early nineteenth century Berlin architect who designed many of the city's museum pieces.

Until around 1948, East Berlin's city planning was modernist, as the galleried apartments that made up the first part of a planned Stalin Allee suggested, but the government condemned these experiments and the rest of the Stalin Allee was designed in accordance with Baroque traditions that were common features of socialist realism, such as symmetry between buildings, the primary street, the urban square and block-perimeter construction. The rest of the avenue was designed by former modernists who had by now ostensibly converted to socialist architectural styles, including Herman Henselmann and Richard Paulick. Henselmann played an important role in the postwar construction of East Berlin's socialist built environment. He designed the famous Frankfurter Tor towers (Figure 7) at the eastern side of the avenue, based on the towers in key government buildings and hotels built a little earlier in Moscow – especially Moscow's 'Seven Sisters' project, which involved creating seven monumental buildings in the wedding cake style across central Moscow. He was also, though, influenced by the domes atop many of the Schinkel-designed buildings in central Berlin. But despite the emphasis in Moscow on local Berlin traditions, it was none the less extremely difficult for German architects to formulate alternative styles within socialist realism.

Figure 7 Karl Marx Allee showing residential buildings and Frankfurter Tor, used by permission of the author

East German planners and architects travelled to Moscow to learn about 'experiences and principles of proper Soviet architecture and planning' (Strobel 2003: 136). What was remarkable about these early visits to Moscow was that it was the Soviet architects, dismayed at the monotonous designs devoid of cultural or social identity or traditions brought by East German architects and planners, who encouraged a more locally sensitive architecture and planning. Strobel (2003: 137) wrote of one delegation: 'The delegation remained in the Soviet Union for over six weeks studying the architectural styles, the construction methods and urban design principles of socialist realism, and together with their instructors they formulated the "Sixteen Principles of Urban Planning", a manifesto that based East Germany's rebuilding program on the Soviet model.' These principles included, for example, a focus on monumentalism in the city centre, a prohibition of garden cities viewed as 'Western', and a focus around central squares and rivers. The principles constituted a regime of truth in so far as they acted as a central reference point for urban learning assemblages. Despite the emphasis in Moscow on local traditions, argued Strobel, in practice German architects had very little space to experiment (on the dissemination of Soviet planning practice to China through exchange programmes, see Wang 2010).

These discussions were, to be sure, instructions sedimented in codified knowledge, and were made policy in the GDR in the mid-1950s with little change. Indeed, the urbanism of East Germany in the 1950s was based on learning as compliance, i.e. on learning codes, with Moscow often heralded as the key example. Architects and planners who seemed to stray away from these principles towards what appeared as modernist planning were reprimanded, often publicly in planning media. It is for this reason that there are such similarities in urban architecture and planning across the former Eastern bloc countries, even though Soviet planners explicitly talked of incorporating local traditions. The rapid speed and thorough influence of socialist realism is all the more remarkable given its short life span – when Krushchev assumed power in Moscow, socialist realism was quickly abandoned in favour of high-rise modernism which was cheaper and quickly constructed. This rapid and extensive urban learning can be attributed to contexts of indoctrination and fear of public exposure for non-compliance (Strobel 2003). The *object* of learning in socialist realism was to create a distinction from Western capitalist modernism through an urban *imaginary* of monumentalism at a human scale. In contrast to Doxiadis' promotion of 'planning as science', the *power* at work in learning here can be described as three-fold: domination, by closing down alternatives despite the nod to local traditions; coercion, through monitoring progress; and inducement, through the use of fear and intimidation to create compliance. The *organizational form* of learning through socialist realism, as with Doxiadis' scientific modernism, was of networks, conferences, study tours and journals. If these urban planning and design examples from the 1950s point to the crucial role of ideology in shaping translocal urban learning assemblages of ideas, people, money and materialities in contexts of deeply unequal power relations, there has none the less been a significant step-change in the mobilities of urban learning, enabled particularly by the Internet and increased international travel. The next section examines some of the politics of contemporary urban policy learning.

Neoliberal Urban Learning Assemblages

If travelling urbanism is a far from new phenomenon, urbanism is none the less increasingly assembled through a variety of sites, people, objects and processes: politicians, policy professionals, consultants, activists, publications and reports, the media, websites, blogs, conferences, peer exchanges, and so on. As McCann and Ward (2010: 175) write:

> As waves of innovation arrive more frequently, a concordant 'churning' has been identified in urban policy, with new ideas and initiatives replacing old with increased regularity. … Contemporary policy-making, at all scales, therefore involves the constant 'scanning' of the policy landscape, via professional publications and reports, the media, websites, blogs, professional contacts, and word of mouth for ready-made, off-the-shelf policies and best practices that can be quickly applied locally.

It is in this context that recent work on urban policy mobilities has emerged. These policy mobilities are translated through local histories and policy contexts in complex ways, from the circulation of revanchist policies through cities like New York and São Paulo, to the travelling free-market ideologies propagated by the World Bank or think-tanks like the Manhattan Institute or Heritage (Stone and Denham 2004; Stone and Maxwell 2004; Peck 2006), to variants of urban entrepreneurialism drawn from seductive 'success' stories like Shanghai and Singapore to Mumbai or Delhi, or the widespread 'sale of community and boutique lifestyles' that accompanies the 'new urbanism' movement for city centre remodelling (Harvey 2008: 32). Sometimes, of course, this is learning only in name and the purpose is to confirm what is already known, or to support existing politico-corporate agendas, while in many cases urban learning may be reduced to a direct or indirect process of imposition or instruction rather than dialogue and reflection.

For policy-makers in many contemporary cities, learning takes place in contexts of development aid and loan agreements with international donors, especially the World Bank. If the Bank has become an increasingly important player in urban development policy debates (Ramsamy 2006), it has also been the single most important international agency in the past

20 years to promote the discourse of 'knowledge for development' (see, for example, the influential 1999 World Development Report, *Knowledge for Development*). In the process, the Bank has been variously accused of promoting knowledge and learning that reflects its own neoliberal prescriptions, being overly concerned with its internal learning strategies rather than learning with and from external partners and sources, excluding other epistemological traditions, cultures of knowledge production, and ways of knowing, reducing learning to a technologically determinist practice driven by information and communication technologies (ICTs), especially the Internet and expensive, high-profile expertise networks like the Global Development Network, and narrowly conceiving learning as a linear process of direct knowledge transfer (Mehta 1999, 2001; King and McGrath 2004; McFarlane 2006a). Much of the World Bank's knowledge and learning initiative to date has involved the aggregation of information and knowledge, a process most starkly represented in the organization's commitment to international best practices. International best practice, enshrined in documents like the World Development Reports, or the UN's *Habitat Agenda* for poor urban environments, reflects the confidence that donors place in claims based on homogenized data gathered from a variety of countries. It is an example of learning through coordination that attempts to governmentalize the truth of urban development by constructing both 'problem' and 'solution', to close down alternatives, and to allow the rapid travelling of a seductive model-as-answer.

Tomlinson (2002) argued that this ideological confidence delimits the scope of debate and potential forms of urban development in policy and planning circles. The result is that a range of alternatives and local contingencies are marginalized in the sorts of urban learning that subsequently takes place. He pointed to a variety of examples of this in the context of urban South Africa, from determined World Bank support of privatization of service delivery, to the lack of consideration of the importance of HIV/Aids in the formulation of urban restructuring plans. He wrote of one Bank 'mission', a central *organizational form* for Bank-promoted learning (2002: 381):

> [T]he Bank mission did not require prior knowledge of the local policy environment. What were missing were the data. Inevitably, a model presumed to apply to all countries and the incorporation of data from other countries as indicative of local conditions produces a homogenizing effect. International best practice becomes self-fulfilling.

In this context, ideology not only shapes the nature of urban policy learning, it all but takes the place of learning, with the exception that local evidence must be located and manipulated through translation in order to support a pre-existing position.

For David Ellerman (2002: 286), former Economic Advisor to the Chief Economist at the World Bank, the Bank is a 'development Church' in which 'new learning at the expense of established Official Views is not encouraged'. He argued that the kinds of local knowledge that can contribute to learning are limited by an adherence in institutions like the World Bank to 'Official Views'. Writing about 'branded knowledge as dogma', Ellerman (ibid.) argued:

> The Church or party model fits perfectly with the standard 'dissemination' or transmission-belt methodology of knowledge-based development assistance. The agency believes it holds the best 'knowledge for development' and is to transmit it to the recipients in the developing world through various forms of aid-baited proselytisation.

This does not mean, of course, that policy-makers or civil society groups simply follow Bank prescriptions, as Kumbaya-Senkwe and Lumambo (2010) show in relation to urban water management in Zambia. The World Bank's preferred option had been to create a single water utility from three, and to increase private control, but local urban state actors resisted and publicly criticized the Bank's claims. Indeed, local policy-makers were able to construct the Bank in the mould of the foreign, colonial actor that was unwilling to listen to local concerns, and raised debate locally about power and who had the right to make decisions. But just as a compromise plan emerged, the Bank – alongside, crucially, state supporters in Zambia – used its financial stick to force through its unpopular plan. Kumbaya-Senkwe and Lumambo's (2010) research reminds us that while ideas like 'privatization' do not necessarily originate with, belong to and disseminate from the Bank, but instead through assemblages that undo any crude conception of local versus foreign knowledge, the Bank none the less plays a crucial ideological role in shaping the sorts of learning that go on in particular places.

As Ramsamy (2006: 171) argued, over the past 30 years the Bank has aggressively promoted a neoliberal ideology of 'lean urbanization' in cities: a reduction in state spending; a commitment to privatization in infrastructure and services; and a dismantling of public sector housing, all in the supposed interests of more effective and efficient services for the urban poor. Ramsamy (2006: 168–9) examined the impact of lean urbanization in Zimbabwe, which during the 1980s and 1990s eroded the urban poor's ability to pay for low-income housing:

> The World Bank's market-oriented approach to low-income urban housing has hardly touched the 7.2 million people who live below the property line in Zimbabwe. The multilateral development agencies' fundamentalist faith in the 'magic of the market' and fiscal austerity has not only failed to offer any viable strategy for dealing with growing inequality, but has actually exacerbated it.

If the marketization of urban services is the *object* of learning here, the *power* at work in learning in World Bank-sponsored urban development often involves domination – the closing down of alternatives – through the production of models and best practices that operate as coordination devices that structure learning amongst policy-makers and offer seductive, ready-made 'solutions'. It is a mode of learning that requires, on the one hand, the imposition of particular World Bank knowledges and, on the other hand, an ignorance of local issues and concerns, and it is incredibly well suited to operating across distance. The faith that externally collated knowledge must be superior to local knowledge resonates with the example of colonial urban learning highlighted earlier, but with the difference that the Bank's sense of possessing superior knowledge resides not in cultural supremacy but in its self-proclaimed existence as a 'knowledge bank' that has *already learnt* established 'international best practices' by systematizing knowledge. The *organizational form* of short-term projects facilitates the Bank's influence, the aim of which is the realization of a neoliberal *imaginary* of the competitive, efficient, capitalist city.

But it is not, of course, just the World Bank that peddles forms of urban learning framed through discourses of 'lean urbanization'. For example, cities in need of rapid recovery – what Schneider and Susser (2003) call 'wounded cities' – are often prey to these rapid neoliberal forms of urban learning. Schneider and Susser's (2003) edited collection shows how large-scale corporate developers target damaged landscapes, often taking advantage of tax breaks. At the same time, in war-ravaged cities like Baghdad, Kabul and Beirut, neoliberal globalization is presented by a range of authorities as a pathway to global 'health' (ibid. 18; Graham 2009). These ideological strategies need to be understood in relation to the role of capitalist cities as sites of creative destruction. As David Harvey (2003: 25–6) has argued of reassembling cities: 'They dance to the capitalist imperative to dismantle the old and give birth to the new as expanding capital accumulation accompanied by new technologies, new forms of organization and rapid influxes of populations (now drawn from all corners of the earth) impose new spatial forms and stresses upon the physical and social landscape.' At stake here is not just a question of how city policy-makers draw upon other cities or debates 'out there', but a question of how the urban problematic is itself framed by dominant ideologies. Particular agendas promulgate certain visions of what creative destruction should achieve, and this process of framing urban problems and solutions is itself an active process of learning urbanism. For example, in his study of the Manhattan Institute's role in the circulation of free-market conservative restructuring of New Orleans following the devastation of Hurricane Katrina in 2002, Jamie Peck (2006: 705) showed how the conservative intelligentsia, following an initial vacuum in which they were defensive and disorientated amongst accusations that they had failed New Orleans' predominantly Black population, aggressively framed the city as in need of free-market reconstruction.

The political efficacy of the peddling of conservative ideologies in the post-Katrina environment rested on their ability to frame urban issues and problems. This occurred through the 'outpouring of conservative commentary editorializing, and detailed policy advocacy' from think-tanks that reframe 'tax cut' as 'tax relief' and welfare retrenchment as 'welfare reform', and which involved the Manhattan Institute reconstructing 'their portable model of urban transformation for the inhospitable setting of New Orleans, advocating what amounts to a shock-therapy program of moral reconstruction [e.g. through faith-based groups advocating 'responsible' communities and nuclear families], social control, and economic discipline [e.g. through spending cuts and corporate contractors]' (Peck 2006: 705–6). This reframing was not, of course, simply imposed by conservative think-tanks, and nor did local policy-makers simply 'draw' upon this conservative urban discourse. Instead, the confluence of a weakened city in desperate need of US federal funding, the aggressive conservativism of the Bush White House, and the persistence of think-tanks and right-wing mainstream media in narrativizing a particular urban 'truth' – the need for free-market reconstruction in what was once, in US terms, a left-of-centre city – as necessary and inevitable, constituted a *seductive* and *manipulative* neoliberal urban learning assemblage that would face little resistance. In this example, urban learning is truncated as the framing of a problem through a particular urban truth that proves seductive to a city polity in a desperate position. If the *object* of learning here was the wounded city in need of urgent reform, the *imaginary* at work in that reform was one of fiscal and moral 'responsibility' and control. As with the World Bank discussion earlier, learning proceeds through an exclusionary dialectic of knowing – i.e. the claim that a certain network of groups has the 'answer' – and ignorance of local desires and knowledge.

If we look more widely at debates on the future of urban development, we see that increasingly *learning itself* has become a seductive focus of debate. For example, the last few years have seen numerous examples of cities designated as 'knowledge cities' (Castells 1996), 'creative cities' (Florida 2002, 2005; Peck 2005), or 'smart cities' (see Hollands 2008). If these discourses have received a staggering amount of attention from urban policy-makers, it is in large part because their focus on and accommodation within existing discourses of urban entrepreneurialism, gentrification and privatization means that they are ideologically appropriate to the times. Reflecting on the 'seductive' power of Richard Florida's creative city thesis for 'civic leaders around the world', Peck (2005: 740) noted how 'from Singapore to London, Dublin to Auckland, Memphis to Amsterdam; indeed, all the way to Providence, RI, and Green Bay, WI, cities have paid handsomely to hear about the new credo of creativity, to learn how to attract and nurture creative workers, and to evaluate the latest "hipsterization strategies" of established creative capitals like Austin, TX or wannabes like

Tampa Bay'. Richard Florida's influential thesis, argued Peck (2005: 767), casts urban competitiveness as cultural and economic 'creativity', and constitutes a seductive narrative for policy-makers in an increasingly busy, 'fast policy market'. He argued that these strategies empower, if only precariously (ibid.: 767–8):

> [U]nstable networks of elite actors, whose strategies represent aspirant attempts to realize in concrete form the seductive 'traveling truths' of the creativity script; they give license to ostensibly portable technocratic routines and replicable policy practices that are easily disembedded and deterritorialized from their centers of production ... they legitimate new urban development models and messages, which travel with great speed through interlocal policy networks, facilitated by a sprawling complex of conferences, web sites, consultants and advocates, policy intermediaries and centers of technocratic translation, the combined function of which is to establish new venues and lubricate new channels for rapid 'policy learning'.

Translocal learning assemblages around 'creative cities' depend upon constant translation through alignments of a range of actors and spaces, including conferences, websites, and policy blueprint documents. Often, individuals, such as 'networked experts' and/or charismatic personalities like Richard Florida, play crucial roles in pushing the travelling of ideas between cities, as critical urban geography has increasingly exposed (Olds 2001; McNeill 2009; Bunnell and Das 2010). For instance, writing about the movement from Kuala Lumpur to Hyderabad of the former's particular brand of IT-driven development, Bunnell and Das (2010) show how Chandrababu Naidu, in his term as Chief Minister of Andhra Pradesh, oversaw the development of an IT Park (which later became 'Cyberabad') and invited the Prime Minister of India to inaugurate Cyber Towers, a 10-storey high-tech building, as part of the HITECH (Hyderabad Information Technology Engineering Consultancy) city. They argue that the personality and charisma of Chandrababu Naidu played an important part in Hyderabad's global visibility: 'Apart from becoming the "West's favorite Indian" (Monbiot, 2004), the "poster boy" for the World Bank in India (Sainath, 2004), and a rising reputation as an "entrepreneurial" and "tech-savvy" chief minister, Chandrababu Naidu attended international events such as the World Economic Forum in Davos in 1999 and 2000 (Naidu and Ninan, 2000), and succeeded in attracting high-profile visitors including Bill Clinton and Tony Blair as well as investment' (2010: 280). While they caution against being seduced by very power of personality, 'there is no doubt that the political celebrity of Chandrababu Naidu was very much part of the rise of Hyderabad-as-model (as Mahathir had been for KL/Malaysia, albeit often in much more geopolitically fractious ways)' (ibid.).

For Peck, the success of Florida's travelling learning strategies lie in part in his promotional and presentational skills, but must be viewed in the context of a broader agenda identified by Harvey (1989b) as 'urban entrepreneurialism', reflected in the upsurge in the number of consultancies offering advice to civic leaders on promoting an urban environment of spectacle, play and gentrification to tourists, investors and young 'creatives'. Urban entrepreneurialism signals the proliferation of public–private growth coalitions that focus investment on particular parts and groups in the city over others, and that promote these specific urban areas to investors and tourists through competitive urban place marketing. In this context, Peck argued, strategies to nurture a 'creative' class of young innovative learners are effectively the latest incarnation of gentrification, in this case 'third wave gentrification' in which the local state assumes an increasingly active role in remaking the city for the middle classes (Smith 2002; Lees et al. 2008). The promotion of so-called 'creative classes' or intelligent/smart cities, then, is part of a broader set of debates focused around a range of high-end investments, from ICTs, cultural industries and science and technology parks, to Southampton's claim to be the UK's first 'smart city' by virtue of its multi-application smartcard, Bangalore's claim to be India's 'Silicon Valley', Singapore's IT2000 plan to create an 'intelligent island' transforming work, life and play, and Dubai's spectacular corporate and tourist-driven fantasy environments of 'imagineered urbanism' driven by strangled urban democratic accountability combined with bloated and corrupt financial flows (Kimninos 2002; Davis 2008; Davis and Monk 2007; Hollands 2008).

The increasingly pervasive discourse of 'smart cities' – conflated as it is with multiple different processes relating to cyber, digital, wired, e-cities, and knowledge cities, reflects two neoliberal choices subscribed to by states. Firstly, as Hollands (2008) argues, an underlying element of many self-designated smart cities is an emphasis on business-led urban development, and as such amounts to a recasting of Harvey's (1989b) discussion of urban entrepreneurialism. For example, one self-declared 'smart city' – San Diego, 'Technology's Perfect Climate' – boasts one of the most competitive sales tax rates in California (7.75%) and has a business tax rate lower than any of the largest 20 cities in the US (Hollands 2008: 308). Instrumental and seductive forms of *power* are at work in these discourses, often in the form of a technological determinism that assumes that greater connectivity to online information necessarily leads to more informed, and thereby economically more 'useful', citizens. Secondly, the elitist biopolitical choice of smart city discourse is clear, for instance in the promotion of 'social capital' – citing the role of networks of trust and reciprocity – or 'lifelong learning'. As Coe et al. (2000: 13, cited in Hollands 2008: 309; my emphasis) wrote, smart cities are 'not possible outside of the development of smart communities – communities that have *learned how to learn*, adapt and innovate'. As Bunnell and Coe (2005: 841) argued in relation to Cyberjaya, part of Malaysia's elite Multimedia Super Corridor development (and see Easterling 2005):

> We suggest that buying into the cybercorridor can in part be understood in terms of participating in modes of (self-)government based on lifestyle choice through practices of consumption. 'Intelligent investment' (as one advertisement put it) in a 'dot.com property' in Cyberjaya is thus not just about buying real estate – though it definitely is partly that – but is also about investing in oneself and one's family for a supposedly immanent information age.

As critical education literature on discourses of lifelong learning has powerfully demonstrated, such arguments promote a self-governing entrepreneurial and – crucially – always unfinished self who must actively and constantly develop new knowledge and skills that can be put to work in a flexible, precarious labour force (e.g. Popkewitz *et al.* 2006; Simons 2006; Fejes and Nicoll 2008; Olssen 2008).

Discourses of smart/creative cities increasingly circulate translocally, whether through conferences of policy-makers and consultants or urban study tours, or through an emerging informed consensus about what the competitive city looks like – fast, technological, research-intensive, aesthetically highly ordered and manicured – constituting both the *object* and *imaginary* of learning. If this consensus is, to be sure, differentiated across cities, it none the less reflects a broad-ranging *neoliberalism of translocal urban learning* as explicitly about public–private investments in particular areas and groups within the city over others in a context of welfare reduction. This neoliberal logic – a competitive-market ideology of privatization, decollectivization, state welfare reduction, and urban public–private entrepreneurialism (Harvey 1989b, 2005) – is informed by an expanding global consultariat and by pervasive élite urban imaginaries, and is manifested in exclusive, segregated urban forms. The spatial formations and politics of these trends take the shape of gated enclaves (Davis 1990; Caldeira 2000), gentrified neighbourhoods predicated on the displacement of working class communities (Smith 2002; Lees *et al.* 2008), and a punitive urbanism often intolerant of the poor and determined to 'take back' urban space through revanchist seizure (Smith 1996, 1998, 2002; Wacqant 2007, 2008).

For example, Kate Swanson (2007) has argued that in the city of Guayaquil, Ecuador, American-style revanchist policies have been implemented with particular rigour. In 2002, the Municipality of Guayaquil contracted infamous former New York City Police Commissioner William Bratton to help shape the city's urban regeneration strategy. Bratton is well known for co-authoring New York City's Police Strategy No. 5 along with former Mayor Rudolph Giuliani, a document Neil Smith (1998:2) described as the 'founding statement of a fin-de-siècle American revanchism in the urban landscape'. Flown in from the US, Bratton was paid an enormous sum of money for three days of work. He suggested an overhaul of Guayaquil's anti-crime structure, which later became dubbed 'Plan Bratton'. Partly as a consequence, beggars and itinerant vendors can be imprisoned for up to seven days, while fines can be as high as US$500. Considering its vastly different socio-economic and political landscape, it is curious that Ecuador should turn to New York for inspiration. As Smith (2001: 73) noted, one of the dangers of the 'New York model' is exactly that it could become a 'template for a global, postliberal revanchism that may exact revenge against different social groups in different places, doing so with differing intensities and taking quite different forms' (see Macleod 2002).

We need to be careful here, as Smith indicated, not to underestimate the specific histories and conditions through which cities translate circulating revanchisms. Swanson (2007) shows how revanchism has become enrolled in a long-standing racial politics in Ecuador against indigenous groups, a politics of 'whitening'. The project of urban renewal is informed by a discourse of purity and defilement. Ecuador's élite geographical imaginaries dictate that Indians belong in the rural sphere, and through hygienic racism, their presence in the urban sphere is constructed as a potential source of 'contamination' for whites. In addition, unlike many cities in North America and Western Europe, beggars, street children, and precarious workers are not being displaced to build luxury condominiums for Ecuador's middle and upper classes. Rather, they are being displaced to make way for the global tourist class. Revanchism is not being driven by the demand for gentrified housing but rather by a re-orientation of the city to the tourist economy; the exclusive Bratton-informed policy agenda is inhabited in a distinctive but none the less exclusive way in Guayaquil. If the *object* of learning here is a 'safe' and economically buoyant city, the *imaginary* is one of 'clean', attractive and manicured urban environs, such as that around the city's new waterfront. This revanchism is not simply external imposition from New York, but operates through a translocal urban learning assemblage shaped by a range of histories and cultural politics and within which a certain urban model – Bratton's New York – takes on a seductive appeal at a particular moment for a specific powerful group in Guayaquil.

If these examples underline the importance of neoliberal ideologies shaping urban learning, they press home the challenge of trying to identify specifically what it is that is travelling between cities – an idea, a discourse, a seductive urban imaginative geography, a specific set of policies, a gathering momentum or consensus, or an authoritarian brand of capitalism? A central question here is: how might we conceptualize these kinds of mobilities? Do we describe such mobilities as neoliberal ideology taking very different substantive forms, a confluence of neoliberal logic that travels globally with contingent historical manifestations? The categories of 'importer' and 'exporter' become blurred, and in their place emerges a loose urban learning assemblage of discourses, images, local cultural politics, contingently placed actors, well-timed interventions, and a neoliberal consensus around urban change. At work in this broad and loose consensus is an associational power that takes the efficient, neoliberal city as its object of learning, but for which there is no single organizational form. For Loic Wacquant (2008), this neoliberal ideological consensus is plain enough. He described an emerging translocal urban geography driven by the neoliberal penal state's logic of anti-poor racial prejudice and crime paranoia, through which public officials have (2008: 57, 71; emphasis in original):

> [R]aced to adopt measures mimicking those showcased by then-Mayor Rudolph Giuliani in New York City; and politicians have run head over heels to be photographed alongside the living incarnation of penal rigor, William Bratton, latter-day prophet of the virile religion of 'zero tolerance' and pricey globe-trotting 'consultant in urban policing'. ... [N]eighbourhoods of urban relegation – the decaying favela in Brazil, the imploding hyperghetto in the United States, the declining banlieue in France, and the desolate inner city in Scotland or Holland – turn out to be the prime physical and social *space within which the neoliberal penal state is concretely being assembled.*

Wacquant's language points to the seductive and associational power of translating discourses of punitive neoliberal urbanism: policy-makers are portrayed as running 'head over heels' to 'mimic' globe-trotting consultants like Bratton, who presents 'ready-fit' models that can be enrolled through existing anti-poor prejudices. In this reading, urban learning is translated both through specific intermediaries like Bratton, and through a generalized authoritarian consensus amongst cities that punishes and excludes the very marginal classes and ethnic groups upon which urban capitalist accumulation none the less depends. In the context of the urban Brazil that Wacquant (2008) went on to describe in this piece, these trends amount to a deepening of Teresa Caldeira's (2000) influential account of São Paulo as *City of Walls*, in which privatization, enclosures, the policing of boundaries, and a range of distancing devices create a public space fragmented and articulated in terms of rigid separations and high-tech security: 'a space in which inequality is an organising value' (Caldeira 2000: 4). But Wacquant's argument contains two significant differences: firstly, no longer are the poor simply a group to be defended through erecting safe enclosures, they are now a group that must be aggressively pursued in line with an increasingly generalized turn to incarceration; and secondly, this is not a trend that applies to a handful of particular cities, but is instead the dominant logic of global contemporary urbanism. As Mbembe and Nuttall (2004: 365; emphasis in original) wrote, if the city is increasingly fractured around a geography of fortification and enclosures, underpinned by the growing demand for sociospatial insulation and technologies of security and surveillance, 'the *stranger* and the *criminal* now assume, more than ever, greater prominence in most urban imaginations ... most of which reflect the complexities of class, race, generation and ideology' (see Hoffman 2007; Vasudevan et al. 2008).

So what can we say analytically about this ideological consensus? What is clear is that if neoliberal logic often dominates contemporary urban learning assemblages in planning and policy contexts, it does not circulate as a singular, all-encompassing dominant force, but instead operates as a contingent set of translating logics that reposition the problematic of urban redevelopment as a form of active learning. As Aiwha Ong (2007) has argued, neoliberal logic constitutes a 'migratory technology of governing' that interacts with a range of situated circumstances, practices and political rationalities. This is why Ong (2007: 5) found the concept of assemblage useful: stressing not structural hierarchy but the 'asymmetrical unfolding of emerging milieu', where there is a constant play between the actual and the possible – as the example of water management in Zambia indicated – in contexts of deeply unequal power relations. If conceptualizations of network often entail an emphasis on the coming together of elements towards a single goal, assemblage entails a conceptual openness to unexpected outcomes, practices and entanglements of different actors and rationalities (Chapters 1 and 6; and see McGuirk and Dowling 2009).

Neoliberal logic travels through 'emerging economies', argued Ong, as both a technique of administration and as a metaphor. As the World Bank's emphasis on 'knowledge', or as the 'smart city' discourse suggests, neoliberalism is as much about the management of populations through 'political entrepreneurialism' or 'active learned subjects' as it is about 'free trade': 'Neoliberalism's metaphor is knowledge' (Ong 2007: 5). 'Knowledge cities' have become buzzwords, with the goal being the creation and management of élite self-actualizing and entrepreneurial subjects. It is clearly, of course, deeply unequal: the high-tech service industries of India, Malaysia or China are focused on particular zones and urban corridors that enjoy disproportionate investments, services, tax incentives, a focus on 'skilled' citizens and foreign investment, and which are often gated in élite enclaves. This creates zones of exception and a situation of graduated governing, stimulating a range of exclusions and resistances. Ong highlighted, for instance, resistance from Malay Muslims to a state discourse of high-tech education because it is perceived to be benefiting the Chinese minority, an example that shows how neoliberal logic can be translated by reinforcing ethnic governmentality. In this context, I conceive neoliberalism not as a generalized project that becomes localized, but as a situated and loose collection of logics and processes that have a widespread and sometimes profound – but not predetermined – influence on the politics of urban learning.

In the daunting challenge of unpacking what it is that is being circulated, translated and learnt amongst cities, from policy-makers to influential élite organizations and other urban constituencies, a great deal of work is to be done in elucidating the different logics, geographies, and political economies of urban learning. But the brief survey in this chapter of some of the forces driving urban learning assemblages in three different periods does allow us to posit a provisional organizing framework for conceptualizing urban learning in policy and planning contexts. This framework attends to four aspects of translocal urban learning assemblages that seek to elucidate what Ong (2007) called the 'asymmetrical unfolding of emerging milieu'. firstly, the forms of power that promote or structure particular kinds of learning; secondly, the object of learning, i.e. the epistemic problem-spaces that travelling policy and planning creates and addresses; thirdly, the form of learning, i.e. the organizational aspect of learning; and, fourthly, the imaginary at work in learning, i.e. the kind of reassembling that learning seeks or leads to. Table 1 provides a summary based on the examples in this chapter. This framework demands that we think carefully about what a critical geography of urban learning might look like – a question I address in the next chapter. First, however, I want to close this chapter by reflecting on the limits of explaining policy learning through the work of ideological framing.

Ideology and Explanation: Beyond Diffusionist Story-Making

There is, of course, a danger in the argument that ideology plays a central role in shaping translocal urban policy learning. Indeed, as Ward (2006) has argued, part of the potential of urban policy mobility approaches is their ability to open up the 'black-box' of neoliberalization to reveal how it is differently assembled and experienced, and to avoid it being implicitly portrayed as all-encompassing. There is a risk, for instance, that attending to the role of ideology can mask the role of novelty and the unexpected in constituting policy learning assemblages, as Russell Prince (2010: 172) argues: 'A focus on the generalised and particular neoliberal politics of policy transfers can obscure the technical work that goes into making particular transfers conceivable and, more saliently, the moments where a particular policy transfer takes an unexpected turn to become part of an unforeseen and/or novel policy assemblage' (McCann 2008, forthcoming; Larner 2003). As Chapter 3 on SDI argued, we need to remain attentive to how the materialities, contingencies and everyday practices – i.e. the work of learning as dwelling – which may appear mundane and inconsequential in relation to ideology, can be critical to how learning occurs and to the sorts of urbanism and urban politics that emerge.

As a central element in the constitution of urban learning, dwelling features in how policies emerge and are lived. Policy-makers do not just learn policy through translation and coordination, they also *inhabit* policies. This occurs in a variety of ways that can be important to how policy evolves: chats over coffee or lunch, drinks in the bar after policy conferences or

Table 1 A framework for urban policy learning

Context	Object	Form	Power	Imaginary
Colonial planning (e.g. Bombay)	Instilling 'Western' conceptions of the clean city	Comparative learning from British urban planning documents.	Seductive, instrumental, and inducement.	Contaminated city made clean, ordered, modern, and familiar.
Planning-as-science (e.g. Baghdad)	Instilling 'Western' capitalist values	Ekistics: science of urban settlements; conferences, models, plans, journals.	Associational and seductive	Future cities of grid-based modernist communities
Socialist realism (e.g. East Berlin)	Socialist symbolism; alternative to 'Western' modernism	Study trips; reinforcing styles through journals	Domination, coercion and inducement	Monumentalism with classicism; participatory public space
Lean urbanization (e.g. World Bank in Harare)	Marketization of urban services	Short-term projects; Urban 'missions'; Models and 'best practices'	Domination and seduction	Competitive, efficient capitalist city
Reframing 'wounded cities' (e.g. New Orleans)	Reframing of city following urban devastation	Aggressive reframing of urban 'problems' and 'solutions'; state funding and support; consensus-building media and think-tanks	Seduction and manipulation	Fiscal 'responsibility' and control
'Smart/creative cities'	Fast, technological, 'knowledge'-intensive growth	Business-led competitive urban development; technological determinism; social capital, lifelong learning, and self-government	Instrumental and seductive	Fast, technologically-driven city
Revanchism and gentrification (e.g. Guayaquil, São Paulo, New York)	'Safe' tourist and investment friendly growth	Globe-trotting consultants; ready-fit models; policy/planning consensus around neoliberal logics	Seduction and association	Clean, attractive, manicured élite urban environments

presentations from mobile consultants, through friendship or conflict, or through the particular atmospheres that are created by policy discussions. This inhabiting is a material affair. Policy consultants or study tours often involve moving around a city to inspect its infrastructure, economic spaces or sociocultural facilities. The glossy policy documents can become a focus of how debate is structured and how policy is implemented and contested, and the glitzy PowerPoint™ presentations that are the hallmark of consultants like Richard Florida can play a role in facilitating the message. The inhabiting of policy can also be contested in practice: the constant play of the actual and possible, in the context of often intransient ideological positions and deeply unequal power relations, is central to how policy is translated. Policy learning assemblages are as much about events, doing and practice as they are about the resultant policy formation.

For example, John Friedmann (2010: 318) narrates how he became an advisor on urban and regional development to the Chilean government in the 1960s, an experience that made 'social learning central to my writing, seeing planning as a form of praxis, that is, as the constant alternation of knowledge and action, theory and practice'. Friedmann (2010: 316) writes that he and his Chilean counterparts were not simply trying to implement a body of knowledge, but instead were instead trying to put into practice a wide range of regional planning epistemologies through an uncertain, improvised 'learning by doing ... in real time'. This necessitated, reflects Friedmann (2010: 317), a process of 'mutual learning' between theory and practice, planners and practitioners, in an attempt to develop workable solutions. But this mutual learning was not simply about exchanging information and knowledge, but about developing an awareness of different ways of seeing: 'In addition to language, history, politics and geography, we [outside consultants] were obliged to become familiar with the bureaucratic sub-culture within which we worked and, more generally, how to get things done' (ibid.). Friedmann had to learn not just through translating his ideas about regional planning to Chile, but through dwelling and attuning his education to a different way of seeing working contexts, problems and solutions. But these cultures of learning can transform as political ideology shifts. The plans that Friedmann and his colleagues came up with were shelved when the Christian Democratic regime they had been advising lost power in 1970 to Salvador Allende's socialist Popular Unity, and were definitively abandoned with the 1973 US-backed coup that invested General Augusto Pinochet with dictatorial powers.

If ideology is crucial in shaping translocal urban policy learning, the challenge is to ensure that this does not become a force by itself in shaping accounts, thereby displacing the role of 'on the ground' contingencies that shape learning (Ong 2007; McGuirk and Dowling 2009). For example, one risk is a retreat to a geography of diffusion, where the nature of urban policy learning is reduced to an explanatory framework based on ideology such as neoliberalism. In this respect, the value of a learning assemblage approach is to deploy perspectives specifically aimed at offering alternatives to what Jane Jacobs (2006: 13) called 'diffusionist story-making' – for instance, through perspectives like translation. As Jacobs (ibid.) put it:

> Diffusionist models of explanation have a relatively stable thing moving through space and time by way of social effort. Translation, in contrast, brings into view not only the work required for a thing to reach one position from another, but also the multiplicity of add-ons that contribute, often in unpredictable and varying ways, to transportation.

So, to return to the example of housing construction in the discussion of socialist realism earlier in the chapter, a range of materialities are brought into the picture, 'even a seemingly mundane material like cement, that works to ensure that when an idea like a highrise arrives in a place it is seen as something to rally around, build, maintain and live in. It is this work that allows a thing like the highrise to appear as a global form – to have a global effect' (Jacobs 2006: 13; and see A.D. King's (2004) work on the bungalow as global form).

A great deal of the 'how' of mobile policy is refracted through governmentality approaches (e.g. Larner and Le Heron 2002; McCann 2008), i.e. the critical study of how urban policy 'problems' are defined, how 'solutions' are invented, and the mechanisms, techniques and procedures through which political authorities realize their programmes (e.g. Dean 1999; see discussion of governmentality and SDI in Chapter 3). But, as McCann (2008) argued, if critical literature on urban policy mobilities must be concerned with the ways in which 'urban truths' are produced and policy arenas are made thinkable – such as through the different ideological resources discussed in this chapter – this must be accompanied by analysis of the practices, actors, atmospheres and representations that constitute urban policy as mobile. For instance, McCann (2008), in his discussion of how Vancouver's drugs harm reduction strategy has travelled translocally, focuses on the speeches, maps, spreadsheets, models, rankings, and PowerPoint™ presentations that travel in briefcases or through e-mail and are passed around different actors. Journals like the *International Journal of Drug Policy* have become important vehicles for the translation of the Vancouver model (just as Doxiadis' *Ekistics* journal did for his brand of architecture), alongside documentary film, websites, and journalism. Bunnell and Das (2010) trace how technologies of seduction, including documents, statistics and images as well as discourses of 'smart' and 'intelligent' cities or notions of 'leapfrogging', played important roles in Malaysia's attempt to learn a particular model of urban technological hypermodernity from Tokyo during the 1990s:

> The technological utopian language … and, perhaps more importantly, the numbers and tables, graphs and charts, glossy pictures, and digital simulations deployed to visualize the 'multimedia utopia' of Malaysia in 2020

had extremely powerful effects. These were the technologies of seduction through which policy shifts were legitimized and, ultimately, urban spaces and lives transformed with the development of MSC [the Multimedia Super Corridor]. (Bunnell and Das 2010: 281)

These multiple and overlapping technologies, actors and routes coalesce in assemblages through which translocal learning occurs, and inform discussions and ways of thinking in meetings with local policy-makers and incoming experts, in the spaces of professional and activist conferences and forums, and informal conversation in the cars, buses, planes and dinner tables of fact-finding trips and field visits (McCann 2008).

Again, these translocal urban policy assemblages are far from new. As Saunier's (1999, 2001, 2002) work on municipal connections in the early twentieth century revealed, a series of linkages – formal and informal, permanent or ephemeral – have long bound urban entities that are geographically far apart, either in a single country or across countries. His work has been rooted in

> the world of organisations, and in the world of associations of municipalities, their meetings, their internal regulations, their recurrent activities – but also the more flexible circle of relationships embracing such things as the correspondence between two scholars interested in the functioning of municipal government or the activities of specialist journals, not forgetting the eagerness of a particular municipality to consult with its opposite numbers elsewhere about some particular subject or other. (Saunier 2002: 512)

This effort reveals a complex political and technical geography of municipal mobility, from the circulation of legal rules, designs, models, administrators, town councillors or regulations, to discourses of social reform and the work of transnational municipal associations and charitable foundations, and the ways in which the shapes of these connections are restricted by geographical proximity, common language, political affiliations, personal or organizational links among councillors and municipal officers, national political contexts and diplomatic relations. If these urban relationalities were facilitated from the mid-nineteenth century by the coming of the steam engine, the telegraph and new technologies for reproducing pictures and text, many of them were shaped in turn by deeper historical links that extend in some cases beyond the nineteenth and twentieth centuries and into the Middle Ages. As Nick Clarke (forthcoming) argues, the debate in urban geography and elsewhere on urban policy mobilities could gain a great deal from engaging in these historical literatures. For example, writing of Saunier's work, he argues that it complements and extends urban policy mobility debates by directing 'us towards processes of connection, exchange, and circulation; the agency of individuals, municipalities, networks, associations, and organizations; and the way in which urban problems and solutions are generated out of a transnational field of continual struggle, momentary consensus, and attempted universalization' (Clarke forthcoming: 21).

These multiple bits and pieces of information, organization, historical connection, and informal conversation enable the production of particular 'spaces of comparability' (Prince 2010: 171), a key form of translation for translocal policy learning. As the examples from Bombay's colonial sanitation municipal debates reveal, the active making of comparable spaces has always been a central component of the mobility of urban policy knowledges. Peck (2005: 748, 749), writing in relation to mobile discourses of 'creative classes', noted the frequent use by consultants of 'comparative evaluations of peer-city strategies' in their advice to particular civic leaders – comparison especially with San Francisco and Austin, 'places that define the new urban genre' – while McCann (2008) has similarly argued that urban consultancy experts actively use comparison to constitute global spaces of emulation and regimes of urban 'truth' by producing particular forms of representation. Prince (2010) critically examines how forms of statistical and calculative techniques enabled the definition, delineation, and therefore comparability of 'creative industry' policies in New Zealand drawn from the UK. These calculative techniques – for example, measuring the extent and sectoral scope of 'creative industries' – rendered the creative industries sector knowable in New Zealand and comparable with the UK. This comparison helped legitimate the constitution of an economic sector that had not existed previously in the governmental sense. The role of comparison in urban policy learning in these different accounts lies not just in the travelling of information, but in the constitution of a new way of seeing and framing an urban policy problem, domain, and solution. In the case of the creative industry concept in New Zealand, this way of seeing was not simply a product of policy exchanges, but – as in McCann's (2008) examples – a shift in perception facilitated by newspaper articles, televised debates, and staged events.

One of the reasons the assemblage concept is useful here is that it emphasizes not just stability but the work of reassembly and failure. In Prince's (2010: 180) account, the attempt to create a policy assemblage that translated into the New Zealand context broadly failed due to governmental disagreement about its scope, but was reassembled as an economic policy where the 'creative' discourse acted to frame economic innovation itself: 'Like bricks taken from one fallen building and used in the construction of another, the creative industries concept became a constitutive element in another policy assemblage.' For example, a different form of comparison that was becoming influential in government circles, this time around 'intelligent interventions' as a form of economic transformation, became entangled in this reassembled creative industry discourse (Prince 2010). Here, multiple trajectories, forms of comparison, ideas, actors, techniques and motivations are aligned and realigned in a changing urban learning assemblage. As Mbembe (2004: 375) has argued in relation to Johannesburg, this 'temptation of mimicry' – the desire to copy, to learn directly from another urban experience – can itself lead to mimesis, i.e. the capacity to 'establish similarities with something else while at the same time inventing something original' (Mbembe 2004: 376). This 'capacity to mime' is common to cities, and names a process through which cities learn indirectly by translating and misreading through comparisons with other urban experiences.

The consequence is that a key task for critical geographies of urban policy learning lies in a focus on genealogies of comparison as a key form of learning through translation. Specifically, this would involve critical attention to how different genealogies of comparison produce particular claims about the present, past and future of the city. A range of important questions emerges here. What practices and discourses shape comparison, and what work does comparison do for how urbanism is learnt and produced? How do comparisons travel? How do comparisons emerge from and help to build ideological consensus, or become contested? We might consider, for instance, how certain cities seek to draw upon particular policy discourses of urban redevelopment in their plans for the city. In my own work in Mumbai, for instance, I have been struck at the performative force in certain – predominantly élite – groups in the city of discourses of transforming Mumbai into a so-called 'world city' of formalized global economic flows and high-end finance, information and transport infrastructures. So, for instance, the controversial *Vision Mumbai: Transforming Mumbai into a World-Class City* policy document, produced by consultancy firm McKinsey and Company and élite NGO Bombay First (2003), became enrolled in a broader set of processes seeking to convert Mumbai into 'Shanghai' – a particular politico-corporate comparative reading of redevelopment for Mumbai reflecting an integrationist comparativism, and which excludes other more inclusive imaginaries of the urban. A crucial question at work here, then, is what is the role of power in producing comparisons, and how is comparison dwelt and contested (or not)?

Conclusion

Recent years have witnessed a growth of interest in urban policy mobilities. As important as this emerging literature is, there have been few attempts to provide a framework for conceptualizing the politics of urban policy mobility. In advancing a conceptual framework attuned to the crucial role of ideology in shaping urban policy learning, I have sought to examine the forms of *power* that structure learning, the *objects* of learning that ideology can produce, the *organizational form* of learning that facilitates the propagation of this ideological learning, and the *imaginaries* that provide a potent visualization of ideological learning. In contrast to the presentism of urban policy mobility debates, I have attempted to show some of the ways in which urban policy learning has occurred historically: in colonial urbanism, in Cold War ideological learning, and in contemporary neoliberal policy mobility. In all the examples highlighted, ideology shapes the nature of urban learning and the reassembling of the city.

We live in a period of 'fast policy transfer' (Peck and Theodore 2001), but we need to be careful not to overestimate this. We cannot assume, for example, that the increased speed of policy mobility necessarily leads to a greater impact. The value in bringing different historical moments together, such as colonial urban policy movement alongside contemporary neoliberal learning, is that it serves to remind us that urbanism has always been relationally produced, and that even if the actors, routes, materialities and speeds are different across space and time, the critical framework deployed in this chapter helps to elucidate the nature of learning at work in these different examples. In many of the examples highlighted in the chapter, there is an important dialectic of learning and ignorance. In colonial comparative learning, Doxiadis' scientific modernism, the World Bank, and the Manhattan Institute, external ideas are, for different reasons, constructed by their agents as superior to local knowledge. In these cases, learning can function as imposition of ideological positions: of the clean, modern, 'Western' city, the rational, modern capitalist city, or of neoliberal urbanism. This imposition is itself a form of learning: a different way of seeing and framing the city and particular urban problematics (as we saw with colonial Bombay, Baghdad, East Berlin and New Orleans). In examining learning through mobile policy and planning, it is crucial that we engage with the changing and contingent ideologies that frame urban learning assemblages without either deciding in advance what those ideologies are or how important they will be, and without ignoring the important role of how policy and planning are dwelt, i.e. produced through multiple mundane everyday spaces and activities. The next chapter builds on this discussion by offering a critical geography of urban learning that considers the specific contributions of the concepts of assemblage and comparison.

Chapter Six

A Critical Geography of Urban Learning

Introduction

Opening the black box of urban learning reveals both how central learning can be to the production of urbanism, and the possibilities of learning as a site of progressive urban politics. The critical purchase of the concept of urban learning is not simply in a call to know more of cities, but to unpack and debate the politics of knowing cities by placing learning explicitly at the heart of the urban agenda. The different examples of urban learning considered in the book so far demand consideration of what a critical urbanism of learning might look like. Critical urban learning involves questioning and antagonizing existing urban knowledges and formulations, and learning alternative formulations. It entails exposing and *un*learning existing dominant arrangements that structure urban learning practices and ideologies, whether in relation to revanchist neoliberalism or gentrification or exclusive pronouncements of the 'smart/creative' city. As Neil Brenner (2009: 199; and see Marcuse 2009) has argued, this aim of unmasking the 'myths, reifications and antimonies that pervade bourgeois forms of knowledge' about capitalism, and offering alternatives, is central to the whole project of critical urban thought. But what might a critical geography of urban learning involve?

As a starting point, we might consider three interrelated aspects. Firstly, it would seek to *evaluate* urban knowledges that are given as inevitable or 'truthful', such as élite claims that the city must neoliberalize, must invest in this area over another area, must privatize urban public services, and must limit the kinds of people who have access to public space or curtail the ways in which those public spaces might be used. In addition to critically examining these claims on their own terms, it would identify both how those claims

---

Learning the City: Knowledge and Translocal Assemblage, First Edition. Colin McFarlane.
© 2011 Colin McFarlane. Published 2011 by Blackwell Publishing Ltd.

serve dominant capitalist interests, and reveal how such dominant forms of learning serve to close down choices. This focus on evaluation as a starting point encompasses the power-object-form-imaginary framework outlined in the previous chapter in relation to policy and planning. Secondly, at the same time as exposing these urban knowledges as neither necessary nor truthful, the task is to present an *alternative* set of urban knowledges, imaginaries, logics and practices that entail learning a new kind of city. As Lefebvre (1967: 172, cited in Marcuse 2009: 193) wrote in relation to rights to the city: 'To the extent that the contours of the future city can be outlined, it could be defined by imagining the reversal of the current situation, by pushing to its limits the converted image of the world upside down'. But if the contours of the future city are to be learnt, we need to consider who is involved in that learning.

This, then, is the third task of a critical geography of urban learning: to identify *who should be involved* in learning new urbanisms, where they might come from, and how that learning should take place in the project of developing a more inclusive, just, sustainable city. This is a methodo-political task that entails experimental forms of learning initiatives (cf. Marcuse 2009, on expose, propose, and politicize in respect to urban planning). One form that such a strategy of critique and alterity through collective participatory urban debate might pursue, then, is that of the urban learning forum discussed in Chapter 4. In urban learning forums like those in Porto Alegre's participatory budgeting programmes – integrated into urban planning not as an appendage but as the very form of planning itself through sustained commitment from the state and pressure from civil society – there is the potential of transformation, and of the emergence of a different kind of city. Here, the progressive possibilities inherent in translating and coordinating different concerns and voices in the city are vividly demonstrated where the possibility of a more equal form of urban dwelling is at stake.

In short, then, a critical urbanism of learning would, firstly, expose and *evaluate* existing urban knowledges and their impacts on different groups and places in the city (building upon the critical discussion of policy and planning in Chapter 5); secondly, put in place processes that enable that learning process to be *democratic* and inclusive of different people with different knowledges from different parts of the city (building on the conception of urban learning forums developed in Chapter 4); and, thirdly, *propose* a more equal, socially just, and ecologically sound form of urbanism through that democratic learning. In building upon this to develop a critical geography of urban learning, this chapter has two aims. First, I will attempt to think through what assemblage offers a critical geography of urban learning. In particular, I consider three forms of analysis and orientations that assemblage brings to a critical geography of urban learning as evaluate–democratize–propose: (a) a descriptive focus on learning as produced through *relations of history and potential*, or the actual and the possible; (b) a reconsideration of *agency* of learning, particularly in relation to distribution and critique; (c) *imaginary*, through the register of cosmopolitan composition. The second aim of the chapter is to consider the question of who is involved in the production of critical geographies of urban learning – the 'democratize' imperative of the evaluate–democratize–propose strategy. I consider this crucial set of issues by examining how we, as critical urban researchers, might learn translocal urbanism through a more multiple and postcolonial framing of cities. In doing so, I consider the important role of comparison in translocal urban learning first discussed in Chapter 1 and then in chapters 4 and 5. Taken together, the discussion of assemblage and postcolonial comparative learning deepen, extend and internationalize the evaluate–democratize–propose strategy for critical urban learning, and offer an important counterpoint to the exclusionary forms of urban learning discussed in Chapter 5. Indeed, while the concepts of assemblage and comparison entail distinct histories, forms and usage in the chapter, thinking them alongside one another is one useful route towards an ethico-political and translocal critical geography of urban learning.

The Actual and the Possible

There is nothing *necessarily* critical about the notion of assemblage, any more than there is anything necessarily critical about notions like capital, labour, space or urbanism, but there is a potentially useful conversation to have when assemblage is brought to bear on urban learning. That said, as Nicholas Tampio (2009) has argued in relation to the Deleuzian tradition of assemblage thinking, for Deleuze the notion of assemblage was always political, for instance in his hope that the left might organize itself in assemblages, or 'constellations of singularities', of cautious, experimental egalitarianism. The concept of 'left assemblages' is a political subjectivity oriented towards the actualization of ideals and the realization of potential:

> A left assemblage can take the form of a political party, a non-governmental organization, an anti-war rally, a school environmental club, a punk rock collective, a campaign to legalize gay marriage, or any loose and provisional material and expressive body that works for freedom and equality. Deleuze envisioned the left as a network of intersecting and conflicting assemblages – a garden rather than a tree. ... Deleuze constructed the concept of assemblages precisely to show how the left could nurture diversity and disagreement. (Tampio 2009: 385, 395)

This emphasis on actualization points to the first contribution of assemblage to a critical geography of urban learning: assemblage's *orientation to urban learning as produced through relations of history and potential*.

This orientation is a method for disclosing the multiple temporalities through which urban learning emerges and might be challenged, and urbanism learnt differently. As Tania Murray Li (2007) has argued, assemblage thinking is concerned with how different spatio-temporal processes are drawn together at a particular conjuncture and made stable through the work of particular powerful actors, but can then be made to disperse or realign through contestation, shifting power relations, or new contexts. Assemblage places emphasis on the depth and potentiality of urban sites and actors in terms of their histories, the labour required to produce them, and their inevitable capacity to exceed the sum of their connections (Chapter 1). Assemblage, as Farías (2009: 15) argues, is 'a double emphasis: on the material, actual and assembled, but also on the emergent, the processual and the multiple'. One example here is McGuirk and Dowling's (2009) argument that the analytic of assemblage offers one possible route for conceiving neoliberalism not as a universal and coherent project, or even as a generalized hegemonic process characterized by local contingencies, but as a loose collection of urban logics and processes that may or may not structure urban change in different places. They seek to conceive urban change through the lens of 'situated assemblages' of different logics, actors, histories, projects and practices that serve not to reify neoliberalism as hegemonic and ascendant, but as one set of possibilities among many. If this is an inherently empirical focus – a call to examine practices 'on the ground' in a way that remains 'open to the practical co-existence of multiple political projects, modes of governance, practices and outcomes generated by and enacted through' specific urban strategies (McGuirk and Dowling 2009: 177) – it is also a focus on how we as urbanists come to learn urban change, and on the hold that terms like 'neoliberalism' have on our critical imaginations. Now, this is not to imply at all that neoliberalism is not a process that is driving radically unequal forms of capital accumulation, labour exploitation, real-estate inequalities, gentrification and dispossession in cities across the globe – it clearly is, and its form and consequences need to be squarely tackled in critical urban thought and action (e.g. Peck and Theodore 2001; Larner 2003; Peck 2003; Wacquant 2007, 2008). The aim is not to take less seriously or ignore the role of neoliberalism, but to understand better how neoliberalism as a dominant political economy and progenitor of polarization and marginalization operates both through its interrelation with multiple other important processes causing inequality in specific cities, but which might exceed neoliberalism itself, and to disavow neoliberalism the potential of defining and delimiting the scope of the possible in the critical imagination.

The contribution of assemblage to critical learning here lies not simply in stressing that the city is reconstructed through processes that exceed neoliberalism, but to 'weaken the defining grip of urban neoliberalism on our theoretical imaginations and on the range of analytical possibility' (McGuirk and Dowling 2009: 184). As Dovey (2010: 16; and see De Landa 2006) has argued, assemblage thinking is an explicit attempt to avoid reductionism and essentialism 'through a concentration on the historic and contingent processes that produce assemblages'. In emphasizing potential through its orientation to assembly, reassembly and constitution, assemblage focuses on the disjunctures between the actual and the possible, between how urban learning is produced and lived and how the city might be learnt in ways that lead to more socially just possibilities.

In this, of course, assemblage thinking echoes much of what Frankfurt School critical theory sought to achieve (Brenner 2009: 203–4) – an emphasis on how bourgeois urban knowledge is produced through history by capitalism and on how urbanism might be more justly reconfigured through political economic alternatives, social movements, and the construction of new forms of knowledge. The relation of history-potential is rooted in the traditions of critical theory, given that critical theory of different hues – from Adorno and Horkheimer (1979) to Marcuse (1972) and Lefebvre (1971) – has always sought to identify the constraints and restrictions that capitalism and culture have historically placed on our capacity to know and learn, while developing possibilities for collective recognition and refusal. The notion of potentiality, for instance, was central to the work of the Situationists in the late 1960s and early 1970s, where the city, as a possible space of collective recognition and reciprocity, became the locus of radical potential (see McDonough's (2009) edited collection of Situationist urban writing). The Situationists' cultural experimentation with *dérive*, or of the architecture of 'unitary urbanism' that explicitly sought to challenge alienating norms of modernist urbanism, were inventive attempts at appropriation of urban space for the desires and needs of the alienated and oppressed by learning different ways of being in the city and producing alternative knowledges (McDonough 2009).

Beyond these traditions, however, assemblage adds a particular emphasis on the *process of reassembling* – assemblage thinking asks us to consider how alternative forms of urban learning might be assembled and actualized. By raising this, I do not suggest that assemblage implies a particular content of alterity, whether socialist or otherwise, but that there is an important concern with the *making* of alternative urban learning. But what is it that might be assembled, and how might that assembling take place? Here, I highlight two orientations of assemblage thinking: firstly, the assembly of the commons, and secondly, the assembling potential of 'generative critique', where assemblage functions as a potentiality of gathering for working towards a form of critique that is constantly generating new associations, knowledges and alternatives. An important part of the response here must be a commitment to assembling alternatives that are produced and held *in common*. Here, as Hardt and Negri (2009) have suggested, assemblage can be read as a form of commoning, of bringing into imagination, debate and realization forms of urban learning that are produced through a participatory inclusion, i.e. through the commons. In Hardt and Negri's (2009) formulation, the commons resonates with their earlier notion of the 'multitude' in that singularities are not required to shed their differences in order to form the commons – a close mirroring of a Deleuzian conception of assemblage (Tampio 2009). Hardt and Negri attempt to extend their elaboration of the multitude by defining the common as that which must be shared in and worked towards. They argue that the commons is a practice of interaction, care and cohabitation. It is not the same as the public because it refers to common culture and knowledge, but neither does it signal a body of content (and see Negri 2006). There is no single body of knowledge that must be learnt as a kind of script for an alternative, more just urban world. Instead, the commons is a process of becoming, a doing that constitutes 'an assemblage of affects or ways of being, which is to say, forms of life – all of which rests on a process of making the common' (Hardt and Negri 2009: 124). As Negri (2006: 67) earlier argued, 'the common is an activity, not a result; it is an assemblage (*agencement*) or an open continuity, not a densification of control'.

The common, then, is a kind of gathering of multiple knowledges and ways of being. It is not a category of sameness, but an 'affirmation of singularities' (Hardt and Negri 2009: 124). For Hardt and Negri (2009: 340), in order to be truly common, the struggle to learn alternative worlds can only occur through 'revolutionary assemblages', i.e. through the parallel coordination of movements as multitude composed of a multiplicity of issues and concerns and not around a singular cause (e.g. that of class, race or gender). The common here is an emergent formation that can only be constructed through a 'cooperative fabric that links together infinite singular activities' (Negri 2006: 71). One political challenge, then, is to counter the accumulation of capital with *an accumulation of the commons*, meaning 'not so much that we have more ideas, more images, more affects, and so forth, but, more important, that our powers and sense increase: our powers to think, to feel, to see, to relate to one another, to love' (Hardt and Negri 2009: 283). This is a project of learning to dwell differently through an inclusive and participatory sharing of knowledge and ideas, and developing new and more just ways of perceiving and living urbanism.

Making the commons is a process of actualizing potential. The potential for the assembling of the commons is especially relevant to the city, which for Hardt and Negri (2009: 25) is 'the space of the common, of people living together, sharing resources, communicating, exchanging goods and ideas'. Hardt and Negri are particularly concerned with how the common features as the basis for biopolitical production, by which they mean not just the natural common of land, minerals, water and gas, but the 'artificial common' of language, image, knowledge, affect, code, habit and practice (ibid.). As a space of encounter – the encounter with alterity – the city produces both positive forms of the common (e.g. local culturally plural spaces) and negative forms of the common (e.g. pollution, traffic or social conflicts around issues as diverse as the use of public parks, libraries and community centres to noise and garbage collection). In the face of the increasing escape from the city of encounter by élite groups into gated enclaves, the politics of assemblage thinking is to emphasize the democratic equality of *assembly itself*, of assembling commonality as a multiple and inclusive process of urban learning. If the commons is a potential possibility of assembly, we need to consider how that process of assembling might take place. Here, I turn to the idea of 'generative critique', i.e. to assemblage's focus on the potentiality of gathering different knowledges, voices and concerns.

Generative critique

Bruno Latour (2004: 225) began a provocative paper on critique by asking: 'What has become of the critical spirit? Has it run out of steam?' Latour's concern was that academic critique, broadly conceived, has failed both to keep up with changes in the world 'out there', and to generate new questions and debates. His concern was not just one of engaging the present, but of critique's orientation towards the present. For example, rather than debunking 'matters of fact' put forward by organizations like the state or the media, Latour encouraged us to develop critical tools that speak about, care for and generate 'matters of concern'. So, one important route for Latour is tracing how matters of concern are produced and maintained.

In a discussion of strands of Heidegger and Whitehead, Latour (2004: 245–6) argued that the notion of 'gathering' offers a new direction for critique, where critique is not 'a flight into the conditions of possibility of a given matter of fact, not the addition of something more human that the inhumane matters of fact would have missed', but a 'multifarious inquiry' that seeks to detect '*how many participants* are gathered in a *thing* to make it exist and to maintain its existence'. By 'thing', Latour was referring to the ways in which 'matters of fact' require, in order to exist, 'matters of concern', i.e. they must mediate and assemble a whole variety of different relations. In other words, the concern here is not just with whether claims to urban knowledge are correct or not – although that is obviously crucial – but with how critical and alternative forms of learning are generated and gathered in creative ways. In this reading, the role of the critic is to participate in the gathering process, meaning that:

> ... the critic is not the one who debunks, but the one who *assembles*. The critic is not the one who lifts the rugs from under the feet of the naïve believers, but the one who offers the participants *arenas in which to gather* ... the one for whom, if something is constructed, then it means it is fragile and thus in great need of care and caution. (Ibid. 246; my emphasis)

Rather than a form of critique that would seek to debunk (i.e. disassociate from and subtract from 'matters of fact'), this is a form that would seek to be more closely associated with its objects by tracing and multiplying the learning relations with those objects.

Now, while I would defend the enduring importance of debunking 'factual' claims as central to a critical geography of urban learning – for instance, to turn to my research in Mumbai, the claim, often widely accepted by particular publics, by the local state that it is people living in 'slums' that are to be blamed for shortages of water because they 'steal' and 'waste' it – Latour none the less opens an important set of issues here in terms of what assemblage might offer the form of urban learning critique. Indeed, one of the reasons the local state in Mumbai can make this 'factual' argument about the so-called 'pilfering' of water by slums is because a multiple set of 'matters of concern' – debates on state capacity, water privatization, and cultures of corruption; questions of rights and citizenship; histories of prejudice; questions about the conditions of water pipes and levels of monsoon rainfall, and so on – are variously rolled into and ignored by influential forces in the constitution of a 'fact'. Debunking this claim is crucially important to be sure, but how might we generate new associations around water? How might we multiply the range of opportunities and spaces in which disparate groups might gather in the constitution of different, more just forms of learning about water settlement? In other words, we might not only debunk such urban inequalities, but rather trace, assemble and thereby generate new forms of association, learning and spaces of political elaboration. One example here might be developing community activist forums that bring together matters of concern ignored by the state – including the state's complicity in informal water economies in informal settlements, or the privatization of water – to generate left assemblages that create new relations. This is a rendering of critique as potential, of assemblage's focus on the possibility of critical learning.

Agency and Critical Learning

The second contribution that assemblage brings to a critical geography of urban learning is its *implications for how we conceive agency, and therefore critical learning*. Agency, as Farías (2009: 15) argues, is an 'emergent capacity of assemblages ... it is the action or the force that leads to one particular enactment of the city', and this force is simultaneously social and material. In approaching agency as an emergent process that is distributed across the social and the material, assemblage thinking requires careful consideration of how different materials might matter for how we learn the city, whether those materials be glossy policy documents, housing and infrastructure materials, placards, banners and picket lines, new and old technologies, software codes, credit instruments, money, commodities, or the material conditions of urban poverty, dispossession and inequality.

This focus on distributed agency across social and material echoes critical theory's longstanding concern with a whole range of capitalism's materializations, including the commodity, the materialization of wealth and poverty through capital accumulation, neo-colonial raw material extraction, gated enclaves, and neoliberalism. We might think, for instance, of David Harvey's (e.g. 2008) brilliant elucidations of how urbanization has played a historically crucial role in absorbing the surplus product that capitalists perpetually produce in search of profits, from the Haussmannization of Second Empire Paris's infrastructure, to the suburbanization of the post-war US, to the rapid urbanization of China over the past 20 years. We can also think here of critical scholarship on the urbanization of socionatures (e.g. Heynen *et al.* 2006). What is significantly different, though, is the particular emphasis that assemblage thinking brings to the agency of the *materials themselves* for critical urban learning. One example here is Bennett's (2010) argument for a 'vital materialism' that seeks to counter the privileging of a specifically *human* agency or politics by emphasizing the agentic contributions of non-human forces in shaping the world. Bennett's effort is to try to comprehend materiality both in relation to and independent of human life, i.e. materiality as a process that sometimes encounters and sometimes exceeds the confines of human life and comprehension. Bennett (2010: 20) theorizes materiality not as a stable and isolated set of objects, but as a *process* of changing relations between humans and non-humans within assemblages, 'as much force as entity, as much energy as matter, as much intensity as extension'. Part of this vital materialism is to examine the shared experiences of people and materials, 'to take a step towards a more ecological sensibility' (Bennett 2010: 10), and to a different rendering of critique.

For example, materialities can act as important agents in learning urban resistance. Jockin Arputham, a high-profile Mumbai activist who founded the Mumbai Slum Dwellers Federation (see Chapter 3), recounted how during the 1980s a range of mundane materials featured as agents of urban activism learning. For example, he recounted how the organization's 'phone bill was very low because we discovered how we could use the public phone for free – by inserting a railway ticket into the receiver'. He continued (Arputham 2008: 333):

> We also learnt how to block the phones of ministers. In the Maharashtra assembly, there were questions asked as to how 30 ministers could have their phones cut at one time. We had designed this in Janata colony, with 100 people assigned one day to go to all the minister's houses. Blocking their phones takes just a simple wire and two stones. It made it sound as if the phone was permanently engaged. We could block all 30 ministers' phones at the same time – simply knowing where they were and shorting out their connections.

Here, railway tickets, wires and stones are quasi-objects (Serres 1974; Serres and Latour 1995) that facilitate learning through coordination amongst activists and the possibilities of resistance. These materials acted as coordination tools that allowed the organization to inhibit communication between officials, thereby bringing an advantage when organizing demonstrations and protests. It is an example of tactical learning (Chapters 2 and 3) that creates an opportunity for resistance by using familiar materials in different ways by translating the use of materials through new assemblages. In addition, as leader of the movement, Jockin was constantly updating himself through the tactic of distributed learning: 'I would be on the public phone – checking to see who was OK, who had escaped, who had been arrested. We had people with bicycles and rickshaws standing by' (ibid.). In distributing agency across the social and the material, assemblage thinking involves attending to how a diverse set of materialities can play multiple roles in the experience and possibilities of urban life.

Importantly, this story opens a larger canvas of materiality that can provide important insight into the nature of urban inequality and political economies. Jockin and his co-activists have been centrally concerned with resisting the demolition of informal settlements by the state and private landlords, often pursued in order to provide for so-called 'higher' economic use. In Mumbai, the state's demolition of informal housing, or of water infrastructure in the context of discriminatory discourses of 'water pilfering', allow us a particular insight into urban political economies in action. For example, the destruction of informal housing to make way for shopping malls or residential complexes tells a vivid and violent story about accumulation, enclosure and displacement, and the inequities of land ownership, gentrification, and commoditization (Harvey 2005, 2007; Vasudevan *et al.* 2008). Or to take a different example, the savage destruction of water pipes by the municipal state in the informal settlement of Rafinagar in northeast Mumbai in the winter of 2009–10 due to so-called 'water scarcity' precipitated the intensification of informal economies where local state officials and private groups sold poor-quality water from water tankers to often desperate residents for a higher price. This is not to claim that focusing on these material reconfigurations and practices tells the whole story of urban inequality and political economies – if indeed any site, perspective or approach can tell the 'whole' story – but that focusing upon these material inequalities and practices within informal settlements can provide more than just a window into inequality and political economy: it can helps us to understand the literal stuff of inequality as well as to appreciate the specific urban forms and processes upon which capitalist development depends. Whether in dispossession, destruction, or the use of materials by activists, assemblage focuses our attention on the everyday life of materialities within informal settlements, on how agency is distributed in relation to those materialities, and on how residents and activists learn to cope with, negotiate, resist and advance their prospects in the city.

There are two implications in particular here for a critical geography of urban learning: the first around methodology and the second around responsibility. In relation to the first, there is a methodological challenge for which accounts such as Bennett's vital materialism can be helpful. Bennett's methodological approach is to take seriously the cultural, linguistic and historical ways in which materials are understood, but not to reduce materiality to these lenses. For Bennett (2010: 17), 'vital materialists will thus try to linger in those moments during which they find themselves fascinated by objects'. Here, the researcher is caught in a tension: between the realization that there is much about the agency of urban materials and the connections they are involved in that exceed our purview, and an effort to none the less attempt to appreciate the agentic force of materials for how we learn urbanism. The guiding aim here is to go beyond the self-evident claim that human life is composed of many material parts, towards an appreciation of these materials as active and to understand the changing role of materials in constituting learning in contexts of daily survival, experience, inequality and possibility. One example here might be a micro-focused ethnography of the ways in which urban materials function not simply as objects but as processes that are learnt in diverse and contingent ways. Here, the aim would be to reveal the changing uses and possibilities that materials shape and allow for learning the city, and which would provide a potentially different lens for linking everyday life, uncertainty and the possibilities of alternative urbanism.

Part of the point here is to say that assemblage thinking is *processual* thinking, that is, the agency of assemblage emerges in process, in bringing different actors together, in their dissolution, contestation and reformulation. As feminist science studies theorist Karen Barad (2007) has argued, agency in this reading is less an attribute or property and more a name for the ongoing reconfiguring of the world. For Barad (2007: 151), materiality is understood not as a fixed substance, but as a 'substance in its intra-active becoming – not a thing but a doing, a congealing of [human and non-human] agency'. As Bennett (2005: 461) argued, assemblage asks us to consider the agency not just of each 'member' of the assemblage, but of the assemblage itself – the milieu, or specific arrangement of things, through which forces and trajectories inhere and transform in a context of 'fluid and intensive generation of potential' (Savage 2009: 171). It is in this specific arrangement of things that assemblage can provide insight into the 'stuff' of urban inequality, and into how people learn to cope and negotiate this inequality.

In distributing agency in this way – and this is a second implication for a critical geography of urban learning of attending to materiality – assemblage troubles at where we assign responsibility when we conduct critique. The discussion of translocal learning in SDI in Chapter 3 revealed some of the important ways in which documents and models shape the nature and politics of learning, including particular conceptions of social change. Bennett (2005, 2010) has argued, in the context of the blackout of the North American power grid in August 2003, that the inherently distributive and multiple nature of agency within such a sociomaterial assemblage casts questions over where responsibility, accountability and the ethico-politics of blame lie. Bennett (2005: 464) framed the problem of the distribution of agency as a binary judgement about where we strategically wish to attribute blame: humans or material agents (e.g. a failure in technology versus a failure in governance or funding). But rather than choosing from one of two binary options, it seems to me that a potential contribution of assemblage thinking to critical urban learning here lies in the particular and often unexpected agency of different materialities. How, for example, do plans emerge, and through which sociomaterial geographies? For the critical urbanist, a focus on agency as distributed through sociomaterial assemblages opens multiple space-times of intervention within urban learning assemblages, where the imperative to act critically is one – to borrow from Barad (2007: 296) – of 'meeting each moment, being alive to the possibilities of becoming', an 'ethical call, an invitation that is written into the very matter of being and becoming'. In its orientation towards thick description of relations of history and potential, and in its conceptual focus on distributing agency across the social and material, assemblage thinking diversifies the range of agents and causes of urban learning, while potentially multiplying the spaces of critical intervention and possibility.

Assemblage and the Critical Learning Imaginary

The third contribution of assemblage to a critical geography or urban learning lies in its particular *imaginary* of urbanism. As an imaginary, assemblage connotes collage, composition and gathering. Here, I want to examine the specific orientation that assemblage can bring to urban critical learning through the issue of urban cosmopolitanism, where the question at stake is whether the imaginary of the urban learning assemblage might allow us to work towards, as a political implication, a progressive cosmopolitan urbanism. Perhaps the closest approximation in the social sciences to the image of compositional mixture is the debate on cosmopolitanism. As a name for a particular kind of translocal relation, cosmopolitanism offers one potentially progressive site for how we conceptualize urban learning assemblages. As David Featherstone (forthcoming; and see Jeffrey and McFarlane 2008) argued in relation to 'subaltern cosmopolitanism', cosmopolitanism is not identitarian in the sense of 'being together', but in the mobile relational sense of 'becoming together'. Cosmopolitanism is a relation of encountering, managing or negotiating difference. In this sense, cosmopolitanism in the city might be described as a kind of 'worldliness' (Chapter 2) that is learnt through the negotiation of difference and as a disposition that emerges through dwelling amongst urban cultural diversity, where dwelling here signals not a regressive exclusionary sensibility deployed in relation to other cultures, but an *ideal*, an openness to and celebration of diversity and translocal connection and togetherness that is to be worked towards (cf. Sandercock 1998, 2004; Harvey 2009).

One reading of cosmopolitanism as a normative political project is the sometimes romanticized discourse of 'one-worldism'. As Alain Badiou (2008: 39) argued, in the simple axiom 'there is only one world' there is a political project of togetherness that affirms a decent standard of living for all. He writes:

> A first consequence is the recognition that all belong to the same world as myself: the African worker I see in the restaurant kitchen, the Moroccan I see digging a hole in the road, the veiled woman looking after children in the park. That is where we reverse the dominant idea of the world united by objects and signs, to make a unity in terms of living, acting beings, here and now. (Badiou 2008)

To echo the discussion of the commons earlier in the chapter, this relational one-worldism is not a homogeneous unity. It is an invocation of a single world where an unlimited set of differences exist, and where these differences do not cast doubt over the unity of the world but are its *principle of existence*. One response, then, to the call for a normative cosmopolitanism is a kind of existential cosmopolitanism that locates a privileged 'I' or 'us' in relation to suffering others in a way that insists upon a decent standard of living regardless of national identity and without understating the crucial role of power, conflict and difference. As Tampio (2009: 393) has argued, this is at the heart of the Deleuzian conception of 'left assemblages', where a key political challenge is to understand that politics needs to be defined not according to what is proximate or distant, but by a commitment to solidarity, connection and equality: 'Deleuze encourages us to situate ourselves mentally in the place of intrinsically different beings.' This practice of situating in a cosmopolitan urbanism is one of learning to progressively negotiate difference. It signals an important translocalism inherent in the practice of assemblage thinking as gathering and composition.

The political consequences of this urban one-worldism is not to somehow eschew the ongoing inequalities of class, gender, race, ethnicity, religion, caste, disability or age that are actively folded into the constitution of urban learning. It is, instead, to demonstrate that we are constitutive parts of those inequalities. The response is not simply an espousal of sympathy, but an attempt to constitute urban learning assemblages that work towards a socially just city. One image of the reassembled just city, then, is a progressive cosmopolitan urbanism that constantly invokes an alternative, more inclusive urban commons based on urban learning as a project of mutual recognition, solidarity and resistance. This image echoes Henri Lefebvre's famous invocation of the right to the city, a right cast not just as material access to urban space, but *a renewed right to urban life*. The right to the city, wrote Lefebvre, 'should modify, concretize and make more practical the rights of the citizen as an urban dweller and user of multiple services. It would affirm, on the one hand, the right of the user to make known their ideas on the space and time of their activities in the urban area; it would also cover the right to the use of the centre, a privileged place, instead of being dispersed and stuck into ghettos (for workers, immigrants, the "marginal" and even for the "privileged")' (1991: 2342, translated in Kofman and Lebas 1996: 3; Harvey 2008; Mitchell 2003). This double affirmation – of both access to the city and active participation of a range of groups in the production of the city as a lived reality – resonates with the image of urban learning assemblage as inclusive cosmopolitanism, and echoes the impetus of urban learning forums discussed in Chapter 4.

It is worth recalling at this point the discussion of agency and the non-human in the previous section, because this offers another sense in which assemblage might function as a critical imaginary in the shape of a cosmopolitan becoming together across difference. As Hinchliffe *et al.* (2005) suggested, cosmopolitanism can be conceived beyond rubrics of inclusion and participation. In their work on urban wildlife and ecologies in Birmingham, they extended cosmopolitan thinking beyond a politics of inclusion and instead attempted to write an ontological project of generating experiments that constitute new collectives and politics. They drew on Stengers' (1997) characterization of cosmopolitics in an effort to 'ecologise the political', an experiment with a politics of changing engagements and ontological struggle in relation to urban wildlife biodiversity (Hinchliffe *et al.* 2005: 650). Part of the 'experiment' here is to put knowledge 'at risk', to attend to the unexpected behaviour of urban wildlife in ways that co-produce new assemblages of knowledge, people and wildlife – 'to allow others, of all shapes, sizes and trajectories, to object to the stories we tell about them, to intervene in our processes as much as we intervene in theirs' (ibid. 655–6).

This cosmopolitical experimentation in urban learning aims to develop not just better, more inclusive representations, but alternative ontologies of human/non-human collectivities that change in process, potentially evoking new possibilities for learning and acting in the city. As an approximation of cosmopolitanism, the assemblage imaginary recalls the concern with rights to the city, but does so through a politics of recognition that has the potential implication of generating new urban knowledges, collectives and ontologies. This question of generating new political collectives echoes the discussion of potentiality and generative critique, and represents one vehicle through which the rights to the city might potentially be realized. In this sense, assemblage extends the rights to the city as a process of learning through composition.

Assemblage, then, offers three interrelated contributions to a critical geography of urban learning that extends the schema of evaluate–democratize–propose: firstly, a focus on relations of history-potential, especially in the actualization of the commons and the potential of generative critique as mechanisms of and objectives of an inclusive, multiple urban learning project; secondly, a reconfiguring of agency to accommodate the role of the material in producing critical urban learning; and thirdly, an image of learning as compositional cosmopolitanism across difference. However, while assemblage offers an orientation to and image of the democratization of learning – for instance in the focus on commoning, gathering and cosmopolitanism – it says little about how critical urbanists generate and gather knowledge as they learn the city. A key issue here relates to the specific geographies through which critical urban learning might be produced. In recent years, a central challenge that has emerged in urban debates is the postcolonial challenge of a progressive, multiple and translocal urban learning.

Postcolonial Urban Learning?

In this section, I consider the question of who is involved in the production of critical geographies of urban learning, that is, the 'democratize' imperative of the evaluate–democratize–propose strategy outlined at the start of the chapter. If we are to develop a critical geography of urban learning, it is important that critical urbanists reflect on what our implicit objects of reference are when we write urbanisms, and that we consider how we might translate other urban experiences, knowledges and theories into a more horizontal comparative field. Central to this task is how we as critical urban researchers learn translocal urbanism. In particular, the challenge here is to think through how we grapple with the multiplicity of different cities, and ways of knowing urbanism, across the global North–South divide. Claims about the nature of the city are often made with cities in the 'North' implicitly in mind, i.e. there is often an implicit, unnoticed slippage between claims to knowledge about certain cities (e.g. New York or London) and claims about 'the city' as an abstract, generalized category – a process Bunnell and Maringanti (2010) call 'metrocentricity'. Urban studies have been surprisingly slow at accounting for how the experience of cities in the 'South' might cause us to rethink urban knowledge and urban theory, and to learn urbanism through a more postcolonial global conversation.

Central to this debate is the problem of comparison, a specific form of learning through translation (Chapters 1, 4 and 5). I use the term comparison here expansively as a name not just for specific method, but as a mode of thought that is central to how we learn alternative urbanisms. Comparison in this sense involves thinking about comparison politically – as a means for situating and contesting existing claims in urban theory, expanding the range of debate, and informing new perspectives (McFarlane 2010). Jennifer Robinson's work, especially her (2006) *Ordinary Cities: Between Modernity and Development,* has been at the heart of efforts to postcolonialize urban studies. As Robinson (2006: 2) stated at the start of that book, 'a postcolonial revisioning of how cities are understood and their futures imagined is long overdue'. She has persuasively argued for a 'cosmopolitan approach' to urban studies, central to which is a straightforward question and imperative: 'How are theoretical approaches changed by considering different cities and different contexts?' (Robinson 2002: 549). I consider comparativism as a *translation strategy* for translocal urban learning – strategy in the sense that experiments in comparative thinking can be useful for, firstly, revealing the assumptions, limits and distinctiveness of particular theoretical or empirical claims, and secondly, for formulating new lines of enquiry and more situated accounts. As a translation strategy of critique and alterity, comparativism depends, in part, on a continuous process of criticism and self-criticism.

In the context of a renewed interest in comparativism (e.g. Oliviera 1996; Dick and Rimmer 1998; Brenner 2001; N. Smith 2002; Roy 2003, 2005; Gulger 2004; King 2004; Robinson 2006; Amelang 2007; Nijman 2007a,b; Huchzermeyer 2002; Kantor and Savitch 2005; Ward 2008, 2009; M.E. Smith 2009; McFarlane 2010), it is all the more important to reflect on the potential of comparison as a strategy for learning postcolonial urbanism, and to consider how a 'relational comparative approach', in Kevin Ward's (2008: 4) phrasing, might be pursued. Responding to this postcolonial challenge is at the heart of a critical geography of urban learning assemblages. Drawing upon postcolonial and development scholarship, I will outline two closely interrelated sets of concern for rethinking comparison as a strategy of critical urban learning: *theory culture* and *ethico-politics.*

Theory culture

At a general level, comparison involves translation not just across different sites, issues or processes, but through and in relation to historical cultures of knowledge production. Theory culture, following Mufti (2005: 475), is 'the *habitus* that regulates "theory" as a discrete set of practices' within and sometimes between specialisms and regions. These 'cultures' are relational, not bounded or territorial, and are often shot through with disagreement and differences. Comparative thinking across theory cultures prompts reflection not just on contrasting spaces or processes, but on the ontological and epistemological framings that inform how the world is being debated, how knowledge is being produced and questioned, and about the purpose of knowledge, research and theory. This means also considering the role of disparate assemblage of institutional actors, from journals to forms and patterns of citation to modes of writing and translation that co-constitute theory cultures. By exploring how different theory cultures debate, for example, the city, politics, infrastructure, modernity, or globalization, or whether these objects are even conceived of as such, there is potential to go beyond simply emphasizing diversity or heterogeneity, to developing more situated forms of learning through comparison.

As Patricia Yaeger (2007) has argued in relation to a metropoetics of urban fiction, the critical challenge here is to locate new epistemic groundings for thinking about cities. She wrote, 'we must go beyond the usual suspects: theorists like Georg Simmel, Walter Benjamin, or even Henri Lefebvre, who have taught us so much about European and American alienation, crowding, flanerie, and absolute space', and pointed towards scholars like Achille Mbembe (2004) and his work on 'superfluity', the wealth created through 'dispensable' workers working in Johannesburg's mines (Yaeger 2007: 12). She goes on: 'How can our ethical and imaginative engagements with others around the world be worked into our scholarly infrastructures? How do we create taxonomies for cities and citizens that are at once off the grid and overly taxonomized?' (ibid. 15). This is an ongoing practical, ethical and intellectual challenge, and one with political potential for the sorts of urbanism that we learn, critically affirm and propose. Robinson (2006: 7), for instance, argued for urban comparison as a basis for 'creative learning'. She demonstrated this in bringing the Chicago School and the Manchester School of African studies into dialogue, showing how new questions and forms of urban learning can be inspired by juxtaposing two 'schools'. But at stake here is not simply the question of *content*, but an ongoing critical reflection on the *structures* through which theory cultures preclude translation or allow for only particular kinds of translation (e.g. as integration rather than critical dialogue). This form of learning is uncertain, not just because of its modesty and provisionality, but because in widening the discursive field, the occurrence of unlikely translations can lead to the increased traversing of unfamiliar and unpredictable terrain. There is an important challenge here in trying to contextualize and understand knowledge from places unfamiliar to the researcher – texts that, for example, may require greater effort for an outsider to grasp (Connell 2007). This kind of learning through translation has much to lose or gain depending upon its approach to dialogue with different theory cultures and constituencies, and this raises a complex set of ethico-political issues.

Ethico-politics

Comparative thinking is not neutral, and efforts to learn between urban theory cultures raises ethical and political considerations (McFarlane 2006c; Jazeel and McFarlane 2007, 2010). For Mufti (2005), in the formulation of comparison there is an at least implicit possibility of transformation in the ways in which objects of discussion and analysis such as urbanism are produced, read and debated. There are echoes here of Said's (1993) notion of 'contrapuntality', a concept developed to analyse the 'Western' cultural archive with an awareness both to the dominated history narrated and the 'other histories' against which it acts. Contrapuntality, for Mufti (2005: 478), 'begins to encode a comparativism yet to come, a global comparativism that is a determinate and concrete response to the hierarchical systems that have dominated cultural life since the colonial era'. Writing in a similar vein, Robinson's (2002: 532) 'cosmopolitan approach' to urban studies highlights two strategies for a critical geography of urban learning. First, there is a need to decolonize Euro-North American knowledges and perspectives by consistently asking, 'How are theoretical approaches changed by considering different cities and different contexts?' (Robinson 2002: 549). This is not an attempt to 'add the Third World city and stir', as if arriving at an all-inclusive picture, but to demonstrate that the view from cities often neglected in theory-generation challenges us to develop new, more situated knowledge of emerging global patterns.

Second, there is a need to learn with, on as close a level playing field as possible, the work of thinkers in different places: 'If a cosmopolitan urban theory is to emerge' wrote Robinson, 'scholars in privileged western environments will need to find responsible and ethical ways to engage with, learn from and promote the ideas of intellectuals in less privileged places' (Robinson 2002: 549–50). This requires a critical epistemic interrogation and reworking, such as that found in Appadurai's (2000) formulation of 'strong internationalisation':

> ['Strong internationalization'] is to imagine and invite a conversation about research in which ... the very elements of this ethic could be the subjects of debate. Scholars from other societies and traditions of inquiry could bring to this debate their own ideas about what counts as new knowledge and what communities of judgement and accountability they might judge to be central in the pursuit of such knowledge. (Appadurai 2000: 14)

This involves an ethical commitment to learning and unlearning through different theory cultures. Drawing on Spivak (1993), McEwan (2003: 384) argued that 'unlearning' involves working hard to gain knowledge of others who occupy those spaces most closed to our privileged view through open-ended conversations. Part of this unlearning involves articulating the Western intellectual's participation in the formation of categories like 'Third World city'. This requires a sensitivity to the relationship between power, authority, positionality and knowledge, but it is a set of problems that cannot simply be acknowledged away; our positions, privileges and ways of seeing cannot be stepped around. Here, critical urban learning depends upon a respectful form of strong internationalization that entails working for new collectives that are reflexive of who starts and controls them, as well as who the members are – i.e. to echo Chapter 4, how their *intensity, openness* and *quality* are actualized. It is an image that opens another set of challenges at the centre of the power-relations of urban learning: *learning how to learn from below* (see Edwards 2001; Spivak 2003).

For Spivak (2003: 52), this involves far more than just learning about other cultures, 'this is imagining yourself, really letting yourself be imagined (experience that impossibility) without guarantees, by and in another culture'. Here, learning in new collectives is an ethical imperative that outlines a horizon of transformation: towards a postcolonial project as an ethico-politics of becoming, emphasizing the processual and anticipatory – in McEwan's (2003: 349) words, 'recognising a condition that does not yet exist, but working nevertheless to bring that about'. Developing these sorts of collectives in research around urbanism can contribute to alternative ways of learning and imagining not just urbanism, but the research enterprise itself. This involves asking not just how 'others' see cities, but how they see the world (in regional terms and in other ways) – in short, 'how does the world look … from other locations?' (Appadurai 2000: 8; see Ferguson 2006).

The notion of strong internationalization involves a particular and reflexive engagement with, for example, activist-intellectual forms of knowledge about the lives of people living in informal settlements (see Chapter 3 on SDI), or regimes of academic knowledge production formulated through distinct patterns of collection, citation or judgement. There is a challenge here for urban researchers to connect more closely and more frequently with the worlds, languages and vocabularies of disparate scholars and activists (Desbiends and Ruddick 2006; Jazeel 2007). These ethico-political questions point to a range of practical challenges in working across different theory cultures, including negotiating new forms of collaboration for comparative research through journals, refereeing and editorships, or in supporting scholarship and writing from different contexts. Other examples of strong internationalization might include developing personal contacts and resources, such as a fund for translations to deal with language barriers, universities and departments investing in graduate language skills, creating funds for exchanges, and encouraging more collaborative postgraduate programmes.

As Mbembe and Nuttall (2004: 352) argued, this involves approaching anew, as much as is possible, the city as archive, and 'drawing on new critical pedagogies – pedagogies of writing, talking, seeing, walking, telling, hearing, drawing, making – each of which pairs the subject and the object in novel ways to enliven the relationship between them and to better express life in motion' – gestures of '*defamiliarization*'. These issues present, of course, a range of difficulties. Who sets the conditions for engagement? How do we avoid these kinds of efforts becoming about 'us' being cosmopolitan, rather than a kind of horizontal dialogue? We need to remain mindful of the unequal resources in which different academic contexts are set, as well as of different research traditions, and the demand that well-meaning Western preoccupations might place on them. For example, how realistic is it for academics operating in the context of scarce personal and/or institutional resources (in South and North) to be 'cosmopolitan'? Well-meaning efforts at producing a postcolonial urban studies more situated in its claims and committed to translocal urban learning needs to explore these possibilities tentatively and through genuine discussion and exploration. In these senses, comparativism is fundamentally about power: the epistemic and institutionalized relations of power between different scholarly and non-scholarly communities within and between different cultures of knowledge production, and this entails problematizing the relationship between 'academic' and 'community'. What 'communities' do we write for? How do we constitute 'communities' as we perform urban studies? How can we generate 'networks of cooperation' in urban studies that 'run around and across the periphery'? (Connell 2007: 228). These questions address what a critical urban geography of learning might practically involve.

These two overlapping areas of theory culture and ethico-politics matter if comparison is to function within critical geographies of urban learning. This outline of translational comparison as a strategy has the potential to lead to the collaborative formation of objects and processes of comparison, rather than a privileging of one context over another, and seeks – under constraints of history, positionality and unequal power relations – not to assume that one theory culture represents a norm or standard of knowledge or ways of knowing over others. It seeks to expand the field of inquiry and develop new critical forms and practices of learning that would work to democratize knowledge in urban debates.

Conclusion

At first sight, a critical geography of urban learning based around a broad strategy of evaluate–democratize–propose appears as a familiar statement of critical urbanism. In this chapter I have sought to think through how that strategy might be extended, deepened and made translocal. I have attempted this in two ways: firstly by considering what a key concept in the book – assemblage – might offer a critical geography of urban learning; and secondly by thinking through how the project of critical urban learning might operate translocally across the global North–South divide as a basis for learning a critical postcolonial urbanism, which necessitates a rethinking of the notion of comparison. These two concepts – assemblage and comparison – while distinct in their form and use, need to be thought together if our aim is to develop an ethico-political translocal project of critical urban learning.

To summarize, assemblage offers three interrelated contributions to a critical geography of urban learning that deepens and extends the schema of evaluate–democratize–propose: firstly, a focus on relations of history-potential, especially in the actualization of the commons and the potential of generative learning through gathering different knowledges, orientations, people and voices, as mechanisms of and objectives of an inclusive, multiple urban learning project; secondly, a reconfiguring of agency to accommodate the role of the material in producing critical urban learning, revealing how attending to materiality can provide important insight into urban inequality and political economies of capitalist accumulation, enclosure and dispossession, and to how people learn to cope with or contest these processes; and, thirdly, an image of learning as compositional translocal cosmopolitanism across difference. If assemblage, in its focus on commoning, gathering and cosmopolitanism, offers an orientation to and image of the democratization of learning, we need to look elsewhere for conceptual tools that respond to the postcolonial challenge of a progressive, multiple and translocal urban learning. In particular, a focus upon comparison as a translation strategy of multiplying the knowledges and lines of inquiry through which urbanism is learnt is crucial. One important way in which this might emerge is through a challenging and uncertain ethico-political commitment to sensitively working through multiple theory cultures in the image of strong internationalization. Taken together, assemblage and comparison offer conceptual resources through which a more multiple and inclusive critical geography of urban learning can emerge. As with the long and diverse history of critical urban thought, such a project is an incessant and daunting struggle, but a struggle over the promise, as Harvey (2009: 282) has put it, of making 'tangible the urban geography of our emancipatory dreams'.

Conclusion

If each city is like a game of chess, the day when I have learned the rules, I shall finally possess my empire, even if I shall never succeed in knowing all the cities it contains.

Italo Calvino, *Invisible Cities*
(1997 [1972]: 109)

In this quote from Italo Calvino's brilliant *Invisible Cities*, learning cities is likened to a process of empire-building. To be sure, and as Chapter 5 suggested, there is a long history of learning that is explicitly about empire creation and maintenance, as the story of nineteenth-century European colonial urbanism attests and as contemporary imperialism vividly dramatizes (King 2004; Legg 2007; Graham 2009). In Calvino's book, the quote comes from Kublai Khan and the 'rules' that he refers to are those he searches for in the stories of far-flung cities that Marco Polo regularly brings him. At one point Kublai feels he is 'on the verge of discovering a coherent, harmonious system underlying the infinite deformities and discords' of cities, but when no model suggests itself he gives up (Calvino 1997: 110). Instead, Kublai finds a mirror of cities in the very process of playing chess, allowing him 'to consider each successive state of the board as one of the countless forms that the system of forms assembles and destroys' (ibid.). For him, the empire of knowledge that he wins here is an empty one; there is no blueprint, nothing definitive that can be learnt other than the facile fact that cities are sites of creative destruction. But there is a politics at work in the ceaseless reassembling of a 'system of forms' that so frustrates Khan: a politics of claims to know the city, and to know what is best for the city and its inhabitants.

Learning the City: Knowledge and Translocal Assemblage, First Edition. Colin McFarlane.
© 2011 Colin McFarlane. Published 2011 by Blackwell Publishing Ltd.

Thomas Blom Hansen and Oskar Verkaaik (2009: 8) have argued that cities are characterized by a 'constitutive unknowability', meaning that 'those who claim to know the urban world, those who demonstrate abilities to manoeuvre and control the urban environment, and those who are able to create narratives about the city and about its people … are able to convert the opacity, impenetrability, historicity and latent possibilities of urban life into a *resource in their own self-making*' (ibid.; my emphasis). Those individuals and groups that possess and perform what Hansen and Verkaaik call 'urban infrapower' because they are 'in the know' have not simply plucked these knowledges and dispositions from the ether, but have actively learnt these particular forms and performances of knowing the city (Chapters 2 and 3). Urban infrapower refers not to the 'Rules' with a capital R that Kublai Khan searched for, but to learning urban rules of a different sort: those around how to cope with, negotiate and advance prospects and agendas. The issue here lies not so much in learning more about cities, but in relation to those forms of learning that are intersubjectively valuable. Intersubjectively valuable learning might include, for instance, understanding the tacit rules of reciprocal urban water exchange systems in times of shortage, learning legal knowledges communicated through the work of the Federation of Tenants Association, learning about model house construction or enumeration in SDI's exchanges, or new forms of learning that emerge through urban learning forums or urban policy exchanges.

The critical purchase of the conceptualization of 'urban learning assemblage' lies not in a straightforward call to know more of cities, but to expose, evaluate and democratize the politics of knowing cities by placing learning explicitly at the heart of urban debate. I have attempted to show that learning is critical to how urbanism is produced, lived and contested, and to how we might produce more socially just urbanism. In doing so, I developed a conceptualization of learning assemblage and examined how that conception illuminates our understanding of urbanism. I set out to address five closely related questions in relation to learning and cities: how might learning be conceptualized? How does learning take place on an everyday basis? How does learning occur translocally? How do different environments facilitate or inhibit learning? And how might we develop a critical geography of learning? If the examples and discussion that I offered in response are, to be sure, selective, it is worth reflecting on how far we have come in addressing these questions.

I began in Chapter 1 by developing a theory of learning based on three interrelated ongoing processes: *translation*, or the relational distributions through which learning is produced as a sociomaterial epistemology of displacement and change; *coordination*, or the construction of functional systems that enable learning as a means of coping with complexity, facilitating adaptation, and bringing different forms of knowledge into relation; and *dwelling*, or the education of attention through which learning operates as a way of seeing and inhabiting the world. I used the concept of assemblage to understand the spatialities of urban learning, where space is conceived as a relationality of processual composition where the 'parts' are not atomistic but define the learning assemblage through interaction with one another. Seen through the lens of assemblage, the spaces of urban learning are the product of relations of history and potential, practice and events, and are structured through unequal relations of knowledge, resource and power. The spatiality of the city as learning assemblage emerges as a 'multiplicity of stories-so-far', as 'coeval becomings' (Massey 2005, 2007) co-constituted through relations of near and far, social and material, and distributed through practices of translation, coordination and dwelling. Chapter 1, then, offered a conception of 'learning assemblages' that was then manifested in specific relation to the different urbanisms discussed in the book.

In response to the question of how learning takes place on an everyday basis in the city, Chapter 2 offered a conception of 'incremental learning'. Incremental learning is an experiential geography of urban accretion. It takes a diversity of forms, from the bit-by-bit ways in which informal housing is added to or altered to meet new needs or possibilities, to pooling contacts and resources to develop new economic opportunities, or building reciprocal exchange systems amongst family and friends. Incrementalism is a central process of urban dwelling, an education of perception to what urbanism – in conditions of often extreme inequality, as we saw in the example of street children in Mumbai – might enable and delimit, and to how people might negotiate it. Incremental learning names some of the everyday practicalities that constitute learning assemblages, and the perceptual fields that are both shaped by and in turn alter those practical enactments. In developing this discussion, I tried to show how improvization – contrary to the idea that it is restricted to sudden moments of inspiration – emerges as a particular form of incremental learning that tweaks at and alters existing urban arrangements, and discussed examples as different as constructing informal housing to urban skateboarding. But if incremental learning connotes a gradual temporality of urban change, it is important to reflect upon the multiple spatiotemporal rhythms that inform urban learning experiences, from walking and rhythms of day and night to capital accumulation and urban migration. I went on in the chapter to introduce the notion of tactical learning, a form of resistance that can emerge through everyday dwelling but which might also emerge through translation (e.g. of legal knowledge) and coordination (e.g. of cooperatives, or through figures like Chandrashekhar Prabhu of the Federation of Tenants Association). As the Federation example demonstrated, tactical learning opens alternative possibilities for urban dwelling. Taken together, incremental urbanism, urban rhythm, and tactical learning are selective but important processes through which the everyday city is learnt and contested.

The third question the book set itself was how does learning occur translocally? This was addressed in the discussion in Chapter 3 of SDI's travelling tactics, the examination in Chapter 4 of urban learning forums across the global North–South divide, and the focus in Chapter 5 on translocal urban planning and policy. Importantly, the theory of learning developed in Chapter 1 – of urban learning assemblages constituted through practices of translation, coordination and dwelling – is applicable to all of the contexts of translocal learning in each of these three chapters. Similarly, while the framework for critically engaging with translocal urban policy and planning – focusing on power, object, organizational form and imaginary – was developed in specific relation to the examples discussed in Chapter 5 as a contribution to debates on urban policy mobilities, it is potentially equally applicable to the examples of SDI or urban learning forums. What is specific to how each of the chapters addresses the translocality of urban learning is not a particular theory of translocal learning, but the shape and politics that the specific actors, logics and contexts bring to the form of translocal learning in each case.

Chapter 3, for example, critically examined the nature and politics of SDI's tactical learning, especially through model house construction and enumeration. As translocal urban learning experiments, SDI's travelling tactics reshape political organization in different sites through three sets of practices: firstly, by focusing centrally on a sociomaterial practice of learning in groups as a basis for reimagining housing options; secondly, by creating, through enumerations, an urban geography within settlements that is also a process of learning political organization; and, thirdly, in combination with a certain *realpolitik* of non-party alignment, by contributing to the cultivation of an entrepreneurial conception of the learned citizen negotiating in partnership with the state. Tactics like enumeration or model house construction are examples of tactical learning through coordination devices that are produced and travel through chains of translation. These translocal tactics enable participants to learn forms of political organization, new possibilities for urban dwelling, and ways of negotiating with the state.

The example of translocal urban learning forums in the second half of Chapter 4 examined one particular exchange linked to SDI's work, that between the Indian Alliance and the UK homelessness and low-income tenants group, Groundswell. In contrast to the stated commitment to learning through translation in SDI's exchanges between different cities in the global South, in the exchange with Groundswell – a particular, temporary form of translocal urban learning forum that combined activists and policymakers – members of SPARC, the NGO element of the Alliance, rejected the possibilities of learning from Groundswell because of perceptions of the limited possibilities of learning from organizations based in the global North. As this example revealed, the stubbornly persistent categorizations of global North and South, with their attendant stereotypical connotations, can militate against the prospects of learning through difference. In contrast, other examples from the literature have shown how exchanges can serve to disrupt such geographical categorizations and thereby encourage learning (e.g. Gaventa 1999). In this context, I argued that the conception of learning through translation can be useful if the possibilities of translocal learning are to be enhanced rather than restricted to a reductive conception of learning as direct transfer between 'similar' cities.

Chapter 5 focused the discussion of translocal urban learning assemblages specifically on the domains of planning and policy. If recent years have witnessed a growth of interest in urban policy mobilities, there have been few attempts to provide a framework for conceptualizing the politics of urban policy mobility. In this chapter, I offered a conceptual framework attuned to the crucial role of ideology in shaping translocal urban policy learning. I examined four interrelated domains here: the forms of *power* that structure learning, the *objects* of learning that ideology can produce, the *organizational form* of learning that facilitates the propagation of this ideological learning, and the *imaginaries* that provide a potent visualization of ideological learning. In addition, and in contrast to the presentism of urban policy mobility debates, I attempted to show some of the ways in which translocal urban policy learning has occurred historically: in colonial urbanism, in Cold War ideological learning, and in contemporary neoliberal policy mobility. In all the examples highlighted, ideology shaped the nature of urban learning and the reassembling of the city. Both the Alliance–Groundswell exchange in Chapter 4 and the ideological framing of policy and planning assemblages in Chapter 5 show that translocal urban learning emerges through a dialectic of similarity and difference: that is, the politics of learning in both examples relates to whether learning is reduced to the privileging of particular world-views and objectives, or whether it is opened up to an experiment in learning through difference. The progressive possibilities of learning through translation has been a sustained focus in the book of the potential of translocal urban learning. I took this argument forward in Chapter 6 when discussing the possibilities of rethinking comparison – a particular mode of learning through translation – as a strategy for a postcolonial urban scholarship.

The fourth question the book opened with was around how different environments facilitate or inhibit learning, and was responded to in Chapter 4 on urban learning forums. Urban learning forums entail the possibility of different actors and knowledges within the city coming together to participate, in the context of unequal power relations, in collective learning. If managed carefully to facilitate sustained *intensity, openness* and *quality* (Callon *et al.* 2009), these learning forums take urban planning in uncertain directions and increase the possibility of more socially just urbanism. In urban learning forums such as those in the Porto Alegre participatory budgeting programmes, which are examples of tactical learning that can recast the possibilities of urban dwelling, there is evidence that forums are integrated into urban planning not as an appendage but as the very form of planning itself. In this kind of learning experiment, there is the potential – provided that the state cedes decision-making power to the forum – of the emergence of a different kind of city. As unlikely as these sorts of participatory learning experiments may often seem, we should not allow ourselves to concede to power by becoming cynical about the possibilities of collaborative and dialogic urban learning. However, if urban learning forums are to be active constituents in reassembling the city, there must be commitment year-on-year from the state to the intensity, openness and quality of forums, as well as ongoing pressure and vigilance from civil society. The forum, then, is a specific organized encounter that may be a one-off or part of a series of events, and which emerges from and reshapes learning assemblages. Again, the focus of discussion in this chapter was necessarily selective; there may, of course, be other environments not considered in the book that actively facilitate urban learning assemblages.

The final question that the book started with was how we might develop a critical geography of learning, which was addressed in Chapter 6. In responding, I began with a framework that sought to, first, *evaluate* existing dominant forms of urban knowledge and learning (incorporating the power-object-form-imaginary framework set out in Chapter 5), *democratize* the groups and knowledges involved in urban learning (incorporating the conception of urban learning forums developed in Chapter 4), and to *propose* alternative knowledges and forms of learning based on that democratization. I then went on to consider in the main body of the chapter how that strategy might be extended, deepened and made translocal, and did so in two ways: firstly, by considering what a key concept in the book – assemblage – might offer a critical geography of urban learning; and, secondly, by thinking through how the project of critical urban learning might operate translocally across the global North–South divide as a basis for learning a critical postcolonial urbanism. These two concepts – assemblage and comparison – while distinct in their form and use, were brought together as the chapter sought to develop a translocal geography of critical urban learning.

I argued that assemblage offers three interrelated contributions to a critical geography of urban learning that deepens and extends the schema of evaluate–democratize–propose: firstly, a focus on *relations of history-potential*, especially in the actualization of the commons and the potential of generative critique as mechanisms and objectives of an inclusive, multiple urban learning project; secondly, a reconfiguring of *agency* to accommodate the role of the material in producing critical urban learning, for example in the tactical learning through the coordinating role of mundane materials in the Mumbai Slum Dwellers Federation; and thirdly, an image of learning as compositional translocal cosmopolitanism across difference. If assemblage, in its focus on commoning, gathering and cosmopolitanism, offers an orientation to and image of the democratization of learning, we need, however, to look elsewhere for conceptual tools that respond to the postcolonial challenge of a progressive, multiple and translocal urban learning project. In particular, a focus on *comparison* as a translational strategy of multiplying the knowledges and lines of inquiry through which urbanism is learnt is important. One way in which this might emerge is through a challenging and uncertain ethico-political commitment to sensitively working through multiple theory cultures in the image of what Arjun Appadurai (2000) called 'strong internationalisation'.

In responding to the five questions that I started with, I have attempted in this book to offer a critical geography of urban learning that is attentive to three interrelated domains: firstly, the translocal nature of learning and some of the implications of that translocal learning for policy or activism; secondly, the everyday forms of learning through which people attempt to cope with or advance their opportunities in the city; and thirdly, more organized efforts to foster learning between different urban constituencies. In doing so, my approach has not been to attempt to identify what it is about urbanism that involves learning, but to begin with a conception of what learning is before attempting to locate it – albeit through very specific examples – in the city. For example, *translation* was located in: the learning of urban geographies and political organization through SDI's tactic of enumeration in Chapter 3; the potential of conceiving learning through difference rather than through resort to a category of 'similarity' in translocal urban learning forums in Chapter 4; the ideological entanglements of urban policy as it travels across space in Chapter 5; and in the possibilities of comparative learning for a critical geography of urban learning in Chapter 6. *Coordination* was identified in: the plan for housing cooperatives and in the speech-making of activist Chandrashekhar Prabhu in the tactical learning around legalities in Chapter 2; the use of model house construction in SDI's work in Chapter 3; learning through urban forums in Chapter 4; and in the use of 'best practices' by the World Bank in urban development proposals mentioned in Chapter 5. Finally, *dwelling* was located in: the incremental urbanism crucial to the shape of everyday assemblages within informal settlements in Chapter 2; the knowledges and imaginations that are crucial to the construction of model houses in SDI in Chapter 3; the materialities, contingencies and everyday practices that produce learning through policy mobilities in Chapter 5; and in the critical potential of learning through living with difference in the discussion of cosmopolitanism as an imaginary of learning assemblage in Chapter 6. More importantly, while some instances of learning may be more focused on dwelling and others more focused on coordination or translation, in the different examples of urbanism discussed in the book – of residents living in informal settlements, street children, urban activists, urban planners and policy-makers – an understanding of urban learning can only emerge through thinking translation, coordination and dwelling together.

Urbanism demands learning. If the conception of learning developed in this book is applicable to non-urban contexts, the book has shown that cities – as spaces of encounter and rapid change, of concentrations of political, economic and cultural resources, and of often confusing unknowability – develop particular manifestations of learning assemblages. Cities are distinctive sociospatial formations constituted by a diverse set of specific forms, processes and experiences that change how people know, perceive and inhabit urban space. We can think here of their density, changeability, sometimes overwhelming complexity and illegibility; their role as spaces of tradition, security and history, or of ease, manipulability and sociality; the possibilities they generate for equality, opportunity, struggle, conflict, exploitation and hardship; their spatial sprawl and connection, and their co-locating of multiple sites and experiences, from informal settlements, where most people arriving in contemporary cities will live, to transport or economic corridors, city centres, industrial and service areas, shopping centres, abandoned spaces, parks, subcultures, and often radically varied architecture; their changing rhythms from those of commuters and tourists to schools and nightlife; and their propagation of new politics, lifestyles, imaginaries and technologies. It is through encountering elements of these myriad forms of urban change through the particular lives, contexts, and agendas of policy-makers, activists or residents, that learning the city takes place.

To live in cities is to come up regularly against the unknown. As Marco Polo is at pains to make clear in response to Kublai Khan's insistence on 'learning the rules' of cities in *Invisible Cities* (Calvino 1997: 38, 61): 'Cities, like dreams, are made of desires and fears, even if the thread of their discourse is secret, their rules are absurd, their perspectives deceitful, and everything conceals something else ... a model city ... is a city made only of exceptions, exclusions, incongruities, contradictions'. But urbanites approach the unknown not with a blank slate but with what they already know and in the context of their own lives, resources, plans, hopes, fears, and so on. As Steve Pile (2001: 263) has put it: 'Knowingness and unknowingness are constitutive of the city: each clads buildings in layers of visibility and invisibility, familiarity and surprise'. To think of the city through the concept of urban learning assemblage is to think both in and out of the city, to think both of how learning assemblages matter in the production, daily life and contestation of the city and to recognize that learning the city cannot be reduced to urbanism as a form or process alone.

I want to close the book by identifying three implications of the conception of urban learning assemblages for how we think and research the city. Firstly, the conception of urban learning assemblage demands attention to how learning features in the efforts of different groups to assemble the city. For instance, current debates about the 'creative' or 'smart' city value particular exclusive groups, spaces and forms of urban development, particularly around well-educated élites living in premium residential spaces and working in high-end service economies, including in science, technology, research, media and finance. Through attending to the historical processes of assembling knowledge, the concept of urban learning assemblage can serve to expose which groups and ideologies have the greater capacity to render urbanism in particular ways over others through, in part, privileging particular forms of urban learning. At stake here is the critical relationship between the actual and the possible city, between the city that has been produced and the city that might have been or that might otherwise arise.

Secondly, and following this, one opening or direction for thinking cities through the concept of urban learning assemblage is to do with the idea of *potential*. As I argued in the book, assemblage connotes not just the history of assembling urbanism through learning, but the *potential* of learning to be otherwise. It signifies urban learning not simply as an output or resultant formation, but as ongoing construction. Assemblage connotes the actualizing of alternative urbanisms through learning urbanisms that could be more socially just, but the content of which have to be learnt through a participatory urban culture (e.g. through urban learning forums). In the face of exclusive claims by élite groups to know what is best of the city – those who would advocate, through ideologies of neoliberalism, exclusionary forms of creative, smart or intelligent cities, or gentrified and gated urbanism – and in contexts of often extreme poverty, violence, inequality and exploitation, we need to hold on to the potential of other, more socially just urbanisms. Cities are places of unexpected encounters, progressive ideas, forms of knowledge and activism, and can generate not only inventive ways of perceiving and acting in urban space, but new forms of urban learning and possibility (e.g. through movements like SDI, the Mumbai Slum Dwellers Federation, the Federation of Tenants Association, or through urban learning forums).

As David Harvey (2008: 33) argued, for example, in urban social movements there is the potential to 'reshape the city in a different image from that put forward by the developers'. By forcing different knowledges and perspectives in public spaces, through protest, or in negotiation with state actors, and by embarking on and promoting alternative forms or experiments in urban learning, it is in these movements that much of the drama of urban politics will be played out. These movements are not necessarily aimed at the state, but often inhabit a wider politics of urban learning that aims to raise awareness and understanding of issues as diverse as urban asylum and racism, to international disaster and crises, and threats to public spaces. In these movements and struggles, argued Harvey (2008: 40), the diverse project of 'right to the city', both as a working slogan and political ideal, is continually posed anew through a focus 'on the question of who commands the necessary connection between urbanization and surplus production and use': 'The democratization of that right, and the construction of a broad social movement to enforce its will is imperative if the dispossessed are to take back the control which they have for so long been denied, and if they are to institute new modes of urbanization' (and see Nicholls 2008). This concerted and sustained campaign of rights to the city is a fraught and contested project of learning the forms, contours and imaginaries through which a better city might be assembled.

Thirdly, and finally, one important question that emerges from the book but which has not been examined is how to research the *translocal* in urban learning assemblages? I was reminded of this methodological challenge in the summer of 2009, when I attended part of a Cities Alliance meeting in Berlin that brought together a range of urban activists and specialists from across the world to discuss urban development. The event, entitled 'Local Challenges, Global Opportunities: An Urban Dialogue', included key representatives from the World Bank and UN Habitat as well as leaders from SDI, the Asian Coalition for Housing Rights (ACHR), and various governments. The emphasis of the discussions was on how to facilitate greater participation of the poor in the urban planning process, and to ensure greater 'horizontal' exchange between civic leaders and civil society groups within and between cities. The theme of learning was central to the discussion: questions about how the poor learn when they migrate to informal settlements and how that learning can be supported to enhance their security in relation to skills or access to work and political institutions; debate on how to link the knowledge of the poor with the day-to-day work of urban governance – as one UN Habitat official put it: 'How can we put data into politics?' – discussion on how to learn from existing forms of peer learning within and between cities, and how to promote exchange learning further. Throughout, there was a consensus that learning within and between cities was central to the prospects of a more progressive urbanism. While SDI and ACHR were identified by some as benchmark examples of 'joint learning processes' through which key problems are collectively identified and responses formulated, other important instances of city-to-city learning were highlighted as increasingly common, such as the current effort by Brazilian authorities to provide slum upgrading assistance to Mozambique, based on the São Paulo experience.

As I left the meeting, I reflected on how this sort of translocal urban forum event is increasingly common, especially to urban development debates, and wondered how in researching these sorts of urban learning initiatives we might respond to an array of methodological challenges. For example, how might we design research projects for translocal urban learning forums, especially considering that some convene once a year from more regional groupings that might meet far more regularly? How do we measure the impact of urban forums when they are often plugged into myriad regional networks through which ideas and knowledges are translated? How can we research and construct knowledge about travelling urban ideas and strategies that involve multiple spaces? How do we conduct the sorts of travelling ethnographies that Clifford (1997) has called for, and in doing so, how might we integrate multiple sources from websites, blogs and wikis to grey literature, interviews and participant observation? What are the difficulties and dangers of making generalizations based on specific translocal sites? Or, as Harvey (1996: 23) has put it, 'what is it that constitutes a privileged claim to knowledge and how can we judge, understand, adjudicate, and perhaps negotiate through different knowledges constructed at very different levels of abstraction under radically different material conditions?' And how do we trace the different conceptions, discourses, and political histories of 'learning' at work in such gatherings?

While the history of urbanism is predicated upon the question of *how we might come to know* the city, the question of learning itself has remained black-boxed. This is likely to be due in part to a sense that 'learning' is an incidental process – that the creation and transformation of knowledge and perception is a background story to the central drivers of both urban change and the urban condition. It is also likely to be partly a result of a lingering sense that learning, if not quite *a*political, is somewhat removed from the formation of political struggle and the practice of urban contention. Despite a long history of reference to 'knowledge' and 'learning' in urban debates, learning has not received sustained theoretical or empirical discussion. I have argued, in contrast, that the city, urban research and critical urbanism can benefit through a conceptualization of learning. The concept of urban learning assemblage reveals how a significant part of the struggle over urban life is a struggle of how we come to know and perceive urbanism, whether through everyday experience, activism, or policy-making.

At stake here is a different theorization and lexicon of urbanism that seeks not to displace existing urban theory, but to add to it: the city as a contested learning assemblage, produced in part through practices of translation, coordination and dwelling that constitute an important part of what goes on in social movements, urban participatory forums, policy-making and everyday urban lives. I have argued, for example: that debates on social movements have acknowledged but failed to account for the role of learning in the political strategies, practices, organization and subjectivities of movements; that accounts of urban policy mobilities might benefit from a historical focus on how translocal learning is produced through relations of power-object-form-imaginary; that conceptions of and research on everyday urban life, for instance through incremental practices, improvisation, and multiple rhythms, might be enhanced by an encounter between learning and dwelling; and that debates on critical urbanism might be deepened and extended through a focus on the relations between learning, assemblage and comparison. In short, I have sought, selectively to be sure, to set out how a conceptualization of learning can make important contributions to how we understand urbanism.

References

Abers, R. (2000) *Inventing Local Democracy: Grassroots Politics in Brazil*. Boulder/London: Lynne Rienner Publishers.
Adorno, T. and Horkheimer, M. (1979) *Dialectic of Enlightenment*. London: Verso.
Akrich, M. and Latour, B. (1992) 'A summary of a convenient vocabulary for the semiotics of human and non-human assemblages', in W. Bijker and J. Law (eds), *Shaping Technology/Building Society*. MIT Press: Cambridge, MA, 201–204.
Allen, J. (2003) *Lost Geographies of Power*. Oxford: Blackwell.
Allen, J. (2004) 'The whereabouts of power: politics, government and space', *Geografiska Annaler* 86B: 19–32.
Allen, J. and Cochrane, A. (2007) 'Beyond the territorial fix: regional assemblages, politics and power', *Regional Studies* 41: 1161–1175.
Allen, J. and Cochrane, A. (2010) 'Assemblages of state power: topological shifts in the organization of government and politics', *Antipode* 42: 1071–1089.
Alsayyad, N. and Roy, A. (2006) 'Medieval modernity: on citizenship and urbanism in a global era', *Space and Polity* 10: 1–20.
Amelang, J. (2007) 'Comparing cities: a Barcelona model?' *Urban History* 34: 173–189.
Amin, A (2002) 'Spatialities of globalisation', *Environment and Planning A* 34: 385–399.
Amin, A. (2006) 'The good city', *Urban Studies* 43: 1009–1023.
Amin, A. and Cohendet, P. (2004) *Architectures of Knowledge: Firms, Capabilities and Communities*. Oxford: Oxford University Press.
Amin, A. and Roberts, J. (2008a) 'Knowing in action: beyond communities of practice', *Research Policy* 37: 353–369.
Amin, A. and Roberts, J. (2008b) *Community, Economic Creativity and Organization*. Oxford: Oxford University Press.
Amin, A. and Thrift, N. (2002) *Cities: Reimagining the Urban*. Cambridge: Polity.

Anderson, B. and Harrison, P. (2010) *Taking Place: Non-Representational Theories and Geography*. Farnham: Ashgate.
Anderson, E. (1990) *Streetwise: Race, Class and Change in an Urban Community*. Chicago: University of Chicago Press.
Anjaria, J.S. (2006) 'Street hawkers and public space in Mumbai', *Economic and Political Weekly* (27 May): 2140–2146.
Anzorena, J. (1987) *Training programme by the Society of Area Resource Centres (SPARC): Women Pavement Dwellers and Housing*. Unpublished report.
Appadurai, A. (2000) 'Grassroots globalization and the research imagination', *Public Culture* 12: 1–9.
Appadurai, A. (2002) 'Deep democracy: urban governmentality and the horizon of politics', *Public Culture* 14: 21–47.
Arputham, J. (2008) 'Developing new approaches for people-centred development', *Environment and Urbanization* 20: 319–317.
Asian Coalition for Housing Rights (2000) *Face-to-Face: Notes from the Network on Community Exchange*. Bangkok: Asian Coalition for Housing Rights.
Asian Coalition for Housing Rights (2001) *Housing By People: Newsletter of the Asian Coalition for Housing Rights* 13, June. Bangkok: Asian Coalition for Housing Rights.
Atkinson, R. (1999) 'Discourses of partnership and empowerment in contemporary British urban regeneration', *Urban Studies* 36: 59–72.
Badiou, A. (2008) 'The Communist hypothesis', *New Left Review* 49 (1): 29–42.
Baiocchi, G. (2001) 'Participation, activism, and politics: the Porto Alegre experiment and deliberative democratic theory', *Politics and Society* 29: 43–72.
Baiocchi, G. (2003) 'Radicals in power', in G. Baiocchio (ed.), *Radicals in Power*. London: Zed Books, pp. 1–26.
Banerjee-Guha, S. (2002) 'Shifting cities: urban restructuring in Mumbai', *Economic and Political Weekly* 12 January: 121–128.
Barad, K. (2007) *Meeting the Universe Halfway: Quantum Physics and the Entanglement of Matter and Meaning*. Durham, NC: Duke University Press.
Barry, A. (2001) *Political Machines: Governing a Technological Society*. London: Athlone.
Batliwala, S. and Patel, S. (1985) *A Census as a Participatory Research Exercise: Case Study of SPARC's Pavement Dweller Census, 1985*. Mumbai: SPARC.
Baviskar, A. (2002) 'The politics of the city', *Seminar* 516 (www.india-seminar.com).
Bayat, A. (1997) *Street Politics: Poor People's Movements in Iran*. New York: Columbia University Press.
Benjamin, W. (1999) *The Arcades Project* (trans. H. Eiland and K McGlaughlin). Cambridge, MA: Belknap Press of Harvard University Press.
Benjamin, W. and Lacis, A. (2004) 'Naples', in M. Bullock and M.W. Jennings (eds), *Walter Benjamin: 1913–1926 v. 1: Selected Writings*. Cambridge, MA: Harvard University Press.
Bennett, J. (2005) 'The agency of assemblages and the North American Blackout', *Public Culture* 17: 445–465.
Bennett, J. (2010) *Vibrant Matter: A Political Ecology of Things*. Durham, NC: Duke University Press.

Bingham, N. (1996) 'Object-ions: from technological determinism towards geographies of relations', *Environment and Planning D: Society and Space* 14: 635–657.
Blume, T. (2001) 'Notes from UK on an Exchange trip, Jan 2001', *The Society for the Promotion of Area Resource Centres*. http://www.sparcindia.org, July 2001.
Bolnick, J. (2007) 'Development as reform and counter-reform: paths travelled by Slum/Shack Dwellers International', in A.J. Bebbington, S. Hickey and D. Mitlin (eds) *Can NGOs Make a Difference? The Challenge of Development Alternatives*. London: Zed Books.
Bombay First (2003) *Vision Mumbai: Transforming Mumbai into a World-Class City*. Mumbai: McKinsey & Company/Bombay First.
Borden, I. (2001) 'Another pavement, another beach: skateboarding and the performative critique of architecture', in I. Borden, J. Kerr, J. Rendell and A. Pivaro (eds), *The Unknown City: Contesting Architecture and Social Space*. Cambridge, MA: MIT Press, pp. 178–199.
Borden, I., Kerr, J., Rendell, J. and Pivaro, A. (eds) (2001) *The Unknown City: Contesting Architecture and Social Space*. Cambridge, MA: MIT Press.
Bourdieu, P. (1977) *Outline of a Theory of Practice*. Cambridge: Cambridge University Press.
Bourdieu, P. (1991) *Language and Symbolic Power* (trans. G. Raymond and M. Adamson, ed. J. Thompson). Cambridge, MA: Harvard University Press.
Brenner, N. (2001) 'World city theory, globalization, and the comparative-historical method: reflections on Janet Abu-Lughod's Interpretation of Contemporary Urban Restructuring', *Urban Affairs Review* September: 124–147.
Brenner, N. (2009) 'What is critical urban theory?', *CITY* 13: 195–204.
Bridge, G. (2005) *Reason in the City of Difference: Pragmatism, Communicative Action and Contemporary Urbanism*. London: Routledge.
Briggs, J. and Sharp, J. (2004) 'Indigenous knowledge and development: a postcolonial caution', *Third World Quarterly* 25: 661–676.
Bullock, M. and Jennings, M.W. (2004) *Walter Benjamin: Selected Writings Volume 1, 1913–1926*. Cambridge, MA: Harvard University Press.
Bunnell, T. and Coe, N.M. (2005) 'Re-fragmenting the 'political': globalization, governmentality and Malaysia's multimedia super corridor', *Political Geography* 831–849.
Bunnell, T. and Das, D. (2010) 'Urban pulse: a geography of serial seduction: urban policy transfer from Kuala Lumpur to Hyderabad', *Urban Geography* 31: 277–284.
Bunnell, T. and Maringanti, A. (2010) 'Practicing urban and regional research beyond metrocentricity', *International Journal of Urban and Regional Research* 34: 415–420.
Cabannes, Y. (2004) *Answers to 72 Frequently Asked Questions About Participatory Budgeting*. Quito: UMP-LAC, UN-Habitat, UNDP.
Caldeira, T. (2000) *City of Walls: Crime, Segregation, and Citizenship in São Paulo*. Berkeley: University of California Press.
Caldeira, T. and Holston, J. (2005) 'State and urban space in Brazil: from modernist planning to democratic intervention', in A. Ong and S.J. Collier (eds) *Global Assemblages: Technology, Politics and Ethics as an Anthropological Problem*. Malden, MA: Blackwell, pp. 393–416.

Callon, M., Lascoumes, P. and Barthe, Y. (2009) *Acting in an Uncertain World: An Essay in Technical Democracy*. Cambridge, MA: MIT Press.

Calvino, I. (1997 [1972]) *Invisible Cities*. London: Vintage.

Campbell, T. (2008) 'Learning cities: knowledge, capacity and competitiveness', *Habitat International* 1–7.

Castells, M. (1983) *The City and the Grassroots: A Cross-Cultural Theory of Urban Social Movements*. Berkeley: University of California Press.

Castells, M. (1996) *The Rise of the Network Society*. Oxford: Blackwell.

Chakrabarty, D. (2002) *Habitations of Modernity: Essays in the Wake of Subaltern Studies*. Chicago: University of Chicago Press.

Chatterjee, P. (2004) *The Politics of the Governed: Reflections on Popular Politics in Most of the World*. Delhi: Permanent Black.

Chattopadhyay, S. (2009) 'The art of auto-mobility', *Journal of Material Culture* 14: 107–139.

Chitekwe, B. and Mitlin, D. (2001) 'The urban poor under threat and in struggle: options for urban development in Zimbabwe', *Environment and Urbanization* 13 (2): 85–101.

Cities Alliance (2008) *Annual Report*. http://www.citiesalliance.org/ca/node/12 (accessed December 2010).

Clarke, N. (2009) 'In what sense "spaces of neoliberalism"? The new localism, the new politics of scale, and town twinning', *Political Geography* 28: 496–507.

Clarke, N. (2010) 'Town twinning in Cold-War Britain: (Dis)continuities in twentieth-century municipal internationalism', *Contemporary British History* 24: 173–191.

Clarke, N. (forthcoming) 'Urban policy mobility, anti-politics, and histories of the transnational municipal movement', *Progress in Human Geography*.

Clifford, J. (1997) *Routes: Travel and Translation in the Late Twentieth Century*. Cambridge, MA: Harvard University Press.

Coe, A., Paquet, G. and Roy, J. (2000) 'E-governance and smart communities: a social learning challenge', *Working Paper 53*, Faculty of Administration, University of Ottawa, October.

Collins, H.M. and Evans, R. (2002) 'The third wave of science studies: studies of expertise and experience', *Social Studies of Science* 32: 235–296.

Connell, R. (2007) *Southern Theory: The Global Dynamics of Knowledge in Social Science*. Sydney: Allen and Unwin.

Contu, A. and Willmott, H. (2000) 'Comment on Wenger and Yanow. Knowing in practice: a "delicate flower" in the organizational learning field', *Organization* 2: 269–276.

Conybeare, H. (1852) 'Report on the sanitary state and sanitary requirements of Bombay', *Selections from the Records of the Bombay Government*, 11: New Series. Bombay: Bombay Education Society's Press.

Cooke, B. and Kothari, U. (eds) (2001) *Participation: The New Tyranny*. London: Zed Press.

Cornwall, A. (2000) *Beneficiary, Consumer, Citizen: Perspectives on Participation for Poverty Reduction*. Sida Studies No. 2. Stockholm: Swedish International Development Co-operation Agency.

Couto, C.G. (2003) 'The second time around: Marta Suplicy's PT administration in São Paulo', in G. Baiocchio (ed.), *Radicals in Power*. London: Zed Books, pp. 79–90.

Crang, M. (2001) 'Rhythms of the city: temporalised space and motion', in J. May and N. Thrift (eds), *Timespace: Geographies of Temporality*. London: Routledge, pp. 187–207.

Cullen, J. and Knox, P. (1982) 'The city, the self and urban society', *Transactions of the Institute of British Geographers* 7: 276–291.

Cumbers, A. and MacKinnon, D. (2004) 'Introduction: clusters in urban and regional development', *Urban Studies* 41: 959–969.

Cumbers, A. and MacKinnon, D. (eds) (2006) *Clusters in Urban and Regional Development*. London: Routledge.

Cumbers, A., Routledge, P. and Nativel, C. (2008) 'The entangled geographies of global justice networks', *Progress in Human Geography* 32: 183–201.

Das, R.J. (2000) 'The state–society relation: the case of an antipoverty policy', *Environment and Planning C: Government and Policy* 18: 631–650.

Davis, M. (1990) *City of Quartz: Excavating the Future in Los Angeles*. London: Verso.

Davis, M. (2006) *Planet of Slums*. London: Verso.

Davis, M. (2008) 'Sand, fear and money in Dubai', in M. Davis and D.B. Monk (eds), *Evil Paradises: Dreamworlds of Neoliberalism*. New York: New Press, pp. 48–68.

Davis, M. and Monk, D. (2007) 'Introduction,' in M. Davis and D. Monk (eds), *Evil Paradises: Dreamworlds of Neoliberalism*. New York: New Press.

Debord, G. (1956) 'Les Lèvres nues no. 9', November 1956, 6010, in T. McDonough (ed.) (2009), *The Situationists and the City*. London: Verso, pp. 77–85.

De Certeau, M. (1984) *The Practice of Everyday Life*. Berkeley: University of California Press.

De Certeau, M., Giard, L. and Mayol, P. (1998) *The Practice of Everyday Life. Vol. 2, Living and Cooking*. Minneapolis: University of Minnesota Press.

De Haan, A. and Maxwell, S. (1998) 'Editorial: Poverty and social exclusion in North and South', in A. de Haan and S. Maxwell (eds), *Poverty and Social Exclusion in North and South*. Sussex: Institute of Development Studies, Seminar Series 29(1), pp. 1–9.

De Landa, M. (2006) *A New Philosophy of Society: Assemblage Theory and Social Complexity*. New York: Continuum Press.

Deleuze, G. and Guattari, F. (1981) 'Rhizome', *Ideology and Consciousness* 8: 49–71.

Deleuze, G. and Guattari, F. (1987) *A Thousand Plateaus*. Minneapolis: University of Minnesota Press.

Deleuze, G. and Parnet, C. (2007 [1977]) *Dialogues II*. New York: Columbia University Press.

Desai, V. and Imrie, R. (1998) 'The new managerialism in local governance: North/South dimensions', *Third World Quarterly* 19: 635–650.

Desbiends, C. and Ruddick, S. (2006) 'Guest editorial: speaking of geography: language, power, and the spaces of Anglo-Saxon hegemony', *Environment and Planning D: Society and Space* 24: 1–8.

De Soto, H. (2003) *The Mystery of Capital: Why Capitalism Triumphs in the West and Fails Everywhere Else*. New York: Basic Books.

Dolowitz, D. and Marsh, D. (1996) Who learns what from whom: a review of the policy transfer literature, *Political Studies* 44: 343–357.
Dossal, M. (1991) *Imperial designs and Indian Realities: The Planning of Bombay City, 1845–1875*. New Delhi: Oxford University Press.
Dovey, K. (2010) *Becoming Places: Urbanism/Architecture/Identity/Power*. New York: Routledge.
Dowler, E. (1998) 'Food poverty and food policy', in A. De Haan and S. Maxwell (eds) (1998), *Poverty and Social Exclusion in North and South*. Sussex: Institute of Development Studies, Seminar Series 29:1, pp. 58–65.
Easterling, K. (2005) *Enduring Innocence: Global Architecture and Its Political Masquerades*. Cambridge, MA: MIT Press.
Edwards, M. (2001) 'Global civil society and community exchanges: a different form of movement', *Environment and Urbanization* 13: 145–149.
Edwards, M. and Gaventa, J. (eds) (2001) *Global Citizen Action*. London: Earthscan.
Elden, S. (2001) *Mapping the Present: Heidegger, Foucault and the Project of a Spatial History*. London: Continuum.
Ellerman, D. (2002) 'Should development agencies have Official Views?', *Development in Practice* 12: 285–297.
Evans, M. (1998) 'Behind the rhetoric: the institutional basis of social exclusion and poverty', in A. De Haan and S. Maxwell (eds) (1998), *Poverty and Social Exclusion in North and South*. Sussex: Institute of Development Studies, Seminar Series 29:1, pp. 42–29.
Eyerman, R. and Jamison, A. (1991) *Social Movements: A Cognitive Approach*. Cambridge: Polity.
Farías, I. (2009) 'Introduction: Decentring the object of urban studies', in I. Farías and T. Bender (eds), *Urban Assemblages: How Actor-Network Theory Changes Urban Studies*. London: Routledge.
Farías, I. and Bender, T. (2009) *Urban Assemblages: How Actor-Network Theory Changes Urban Studies*. London: Routledge.
Featherstone, D. (2008) *Resistance, Space and Political Identities: The Making of Counter-Global Networks*. Oxford: Wiley-Blackwell.
Featherstone, D.J. (forthcoming) 'Black internationalism, subaltern cosmopolitanism and the spatial politics of anti-fascism', *Annals of the Association of American Geographers*.
Fejes, A. and Nicoll, K. (eds) (2008) *Foucault and Lifelong Learning: Governing the Subject*. London: Routledge.
Ferguson, J. (1994) *The Anti-Politics Machine: 'Development', Depoliticization, and Bureaucratic Power in Losotho*. Minneapolis: University of Minnesota Press.
Ferguson, J. (1999) *Expectations of Modernity: Myths and Meanings of Urban Life on the Zambian Copperbelt*. Berkeley: University of California Press.
Ferguson, J. (2006) *Global Shadows: Africa in the Neoliberal World Order*. Durham, NC: Duke University Press.
Ferguson, J. and Gupta, A. (2002) 'Spatializing states: toward an ethnography of neoliberal governmentality', *American Ethnologist* 29: 981–1002.
Fischer, G. (2001) 'Communities of interest: learning through the interaction of multiple knowledge systems', in S. Bjornestad, R. Moe, A. Morch and A. Opdahl (eds), *Proceedings of the 24th IRIS Conference, Ulvik*. Bergen: Department of Information Science, Bergen, Norway, pp. 1–14.

Florida, R. (2002) *The Rise of the Creative Class: And How it's Transforming Work, Leisure, Community, and Everyday Life*. New York: Basic Books.
Florida, R. (2005) *Cities and the Creative Class*. New York: Routledge.
Forrester, J. (2000) *The Deliberative Practitioner*. Cambridge, MA: MIT Press.
Foucault, M. (1978) 'Governmentality', in G. Burchell, C. Gordon and P. Miller (eds) (1991), *The Foucault effect: Studies in Governmentality*. London: Harvester Wheatsheaf, pp. 87–104.
Foucault, M. (1980) *Power/Knowledge: Selected Interviews and Other Writings 1972–1977* (ed. C. Gordon). London: Harvester.
Freire, P. (1970) *Pedagogy of the Oppressed*. New York: Continuum.
Friedmann, J. (2010) 'Crossing borders: do planning ideas travel?', in P. Healey and R. Upton (eds), *Crossing Borders: International Exchange and Planning Practices*. London: Routledge, pp. 313–328.
Gandy, M. (2005) 'Cyborg urbanization: complexity and monstrosity in the contemporary city', *International Journal of Urban and Regional Research* 29: 26–49.
Gandy, M. (2006) 'Planning, anti-planning and the infrastructure crisis facing metropolitan Lagos', *Urban Studies* 43: 371–396.
Gaventa, J. (1998) 'Poverty, participation and social exclusion in North and South', in A. De Haan and S. Maxwell (eds), *Poverty and Social Exclusion in North and South*. Sussex: Institute of Development Studies, Seminar Series 29:1, pp. 50–57.
Gaventa, J. (1999) 'Crossing the great divide: building links and learning between NGOs and community-based organizations in North and South', in D. Lewis (ed.), *International Perspectives on Voluntary Action: Reshaping the Third Sector*. London: Earthscan, pp. 21–38.
Gertler, M.S. (2003) 'A cultural economic geography of production: are we learning by doing?', in K. Anderson, M. Domosh, S. Pile and N. Thrift (eds), *The Handbook of Cultural Geography*. London: Sage, pp. 131–146.
Gertler, M.S. (2004) *Manufacturing Culture: The Institutional Geography of Industrial Practice*. Oxford: Oxford University Press.
Gertler, M.S. and Wolfe, D.A. (eds) (2002) *Innovation and Social Learning: Institutional Adaptation in an Era of Technological Change*. Basingstoke: Macmillan/Palgrave.
Gherardi, S. and Nicolini, D. (2000) 'To transfer is to transform: the circulation of safety knowledge', *Organization* 2: 329–348.
Gibson, J. (1979) *The Ecological Approach to Visual Perception*. Boston: Houghton Mifflin.
Gill, L. (2000) *Teetering on the Rim: Global Restructuring, Daily Life, and the Armed Retreat of the Bolivian State*. New York: Columbia University Press.
Glaeser, E.L. (1999) *Learning in Cities*. Cambridge, MA: NBER Working Paper 6271.
GLTN (2006) *Brief Issues Paper in Preparation for the Workshop on Innovative Pro Poor Land Tools*. Nairobi: Global Land Tool Network (GLTN) and UN-HABITAT.
Goldfrank, B. and Schrank, A. (2009) 'Municipal neoliberalism and municipal socialism: urban political economy in Latin America', *International Journal of Urban and Regional Research* 33: 443–462.
Grabher, G. (1993) 'The weakness of strong ties: the lock-in of regional development in the Ruhr area', in G. Grabher (ed.), *The Embedded Firm: On the Socio-Economics of Industrial Networks*. London: Routledge, pp. 255–277.

Grabher, G. (2004) 'Temporary architectures of learning: knowledge governance in project ecologies', *Organization Studies* 25: 1491–1514.
Grabher, G. and Ibhert, O. (2006) 'Bad company? The ambiguity of personal knowledge networks', *Journal of Economic Geography* 6: 251–271.
Graham, S. (2008) 'Robowar™ dreams', *City* 12: 25–49.
Graham, S. (2009) *Cities Under Siege: The New Military Urbanism*. London: Verso.
Graham, S. and Thrift, N. (2007) 'Out of order: understanding repair and maintenance', *Theory, Culture and Society* 24: 1–25.
Greco, J. and Sosa, E. (eds) (1999) *The Blackwell Guide to Epistemology*. Oxford: Blackwell.
Gregory, D. (2000) 'Cultures of travel and spatial formations of knowledge', *Erdkunde* 54: 297–309.
Groundswell (2001a) *India: Diary Report*. www.groundswell.org. July.
Groundswell (2001b) *Exchanges: a Rough Guide*. www.groundswell.org. July.
Gulger, J. (2004) *World Cities Beyond the West: Globalization, Development and Inequality*. Cambridge: Cambridge University Press.
Haas, P.M. (1992) 'Introduction: Epistemic communities and international policy coordination', *International Organization* 46: 1–35.
Habermas, J. (1962, trans. 1989) *The Structural Transformation of the Public Sphere: An Inquiry into a category of Bourgeois Society*. Cambridge: Polity.
Hajer, M. (1995) *The Politics of Environmental Discourse: Ecological Modernization and the Policy Process*. Oxford: Clarendon Press.
Hansen, T.B. and Verkaaik, O. (2009) 'Introduction urban charisma: one everyday mythologies in the city', *Critique of Anthropology* 29: 5–26.
Haraway, D. (1991) *Simians, Cyborgs and Women: The Reinvention of Nature*. London: Free Association Books.
Hardt, M. and Negri, A. (2009) *Commonwealth*. Cambridge, MA: Harvard University Press.
Harris, A. (2008) 'From London to Mumbai and back again: gentrification and public policy in comparative perspective', *Urban Studies* 45: 2407–2428.
Harris, S.R. and Shelswell, N. (2005) 'Moving beyond communities of practice in adult basic education', in D. Barton and K. Tusting (eds), *Beyond Communities of Practice: An Experiential Approach to Knowledge Creation*. Cambridge: Cambridge University Press, pp. 158–179.
Harrison, P. (2007) 'The space between us: opening remarks on the concept of dwelling', *Environment and Planning D: Society and Space* 25: 625–647.
Harvey, D. (1989b) 'From managerialism to entrepreneurialism; the transformation of urban governance in late capitalism', *Geografiska Annaler B* 71: 3–17.
Harvey, D. (1996) *Justice, Nature and the Geography of Difference*. Oxford: Blackwell.
Harvey, D. (1997) 'Contested cities: social process and spatial form', in R.T. LeGates, and F. Stout (2007) (eds), *The City Reader* (4th edn). London and New York: Routledge, pp. 225–232.
Harvey, D. (2003) 'The city as a body politic', in J. Schneider and I. Susser (eds), *Wounded Cities: Deconstruction and Reconstruction in a Globalized World*. New York: Berg, pp. 25–46.
Harvey, D. (2005) *A Brief History of Neoliberalism*. Oxford: Oxford University Press.
Harvey, D. (2007) 'Neoliberalism as creative destruction', *Annals of the American Academy of Political and Social Science* 610: 21–44.

Harvey, D. (2008) 'The right to the city', *New Left Review* 53: 23–40.
Harvey, D. (2009) *Cosmopolitanism and the Geographies of Freedom*. New York: Columbia University Press.
Hayek, F. (1945) 'The use of knowledge in society', *American Economic Review* 35: 519–530.
Headrick, D.R. (1988) *The Tentacles of Progress: Technology Transfer in the Age of Imperialism, 1850 1940*. Oxford: Oxford University Press.
Healey, P. (2002) 'On creating the "city" as a collective resource', *Urban Studies* 39: 1777–1792.
Healey, P. and Upton, R. (eds) (2010) *Crossing Borders: International Exchange and Planning Practices*. London: Routledge.
Heidegger, M. (1971) 'Building dwelling thinking', in A. Hofstadter (ed.), *Poetry, Language, Thought*. New York: Harper and Row.
Hetherington, K. (1998) *Expressions of Identity: Space, Performance, Politics*. London: Sage.
Heynen, N., Kaika, M. and Swyngedouw, S. (eds) (2006) *In the Nature of Cities*. New York: Routledge.
Hinchliffe S. (2003) '"Inhabiting": landscapes and natures', in M. Domosh and S. Pile (eds), *Handbook of Cultural Geography*. London: Sage, pp. 207–225.
Hinchliffe, S., Kearnes, M., Degen, M. and Whatmore, S. (2005) 'Urban wild things: a cosmopolitical experiment', *Environment and Planning D: Society and Space* 23: 643–658.
Hoffman, D. (2007) 'The city as barracks: Freetown, Monrovia, and the organization of violence in postcolonial African cities', *Cultural Anthropology* 22: 400–428.
Hollands, R. (2008) 'Will the real smart city please stand up? Intelligent, progressive or entrepreneurial?', *City* 12: 303–320.
Holloway, J. (2000) 'Institutional geographies of the New Age movement', *Geoforum* 31: 553–565.
Holston, J. (1999) 'Spaces of insurgent citizenship', in J. Holston (ed.), *Cities and Citizenship*. Durham, NC: Duke University Press, pp. 155–176.
Holston, J. (2008) *Insurgent citizenship: disjunctions of democracy and modernity in Brazil*. Princeton: Princeton University Press.
Homeless International (2001) Istanbul +5: Creating a space for all voices? 25 slum dwellers go the UN. Coventry: Homeless International. http://www.theinclusivecity.org/resources/publications/publications.htm, May 2002.
Hosagrahar, J. (2006) *Indigenous Modernities: Negotiating Architecture and Urbanism*. London: Routledge.
Huchzermeyer, M. (1999) 'Current informal settlement intervention in South Africa: four case studies of people-driven initiatives.' Unpublished chapter, Department of Sociology, University of Cape Town.
Huchzermeyer, M. (2009) 'Enumeration as a grassroot tool towards securing tenure in slums: insights from Kisumu, Kenya', *Urban Forum* 20: 271–292.
Hutchins, E. (1996) *Cognition in the Wild*. Cambridge, MA: MIT Press.
Indian Alliance (no date) *Demolitions to Dialogue: Mahila Milan Learning to Talk to its City and Municipality*. Mumbai: SPARC.
Indian Alliance (2008) *Citywatch: A SPARC-NSDF-Mahila Milan Publication*. Mumbai: SPARC.

Ingold, T. (2000) *The Perception of the Environment: Essays in Livelihood, Dwelling and Skill*. London: Routledge.
Ingold, T. (2004) 'Culture on the ground: the world perceived through the feet', *Journal of Material Culture* 9: 315–340.
Ingold, T. (2008) 'Bindings against boundaries: entanglements of life in an open world', *Environment and Planning A* 40: 1796–1810.
Ingold, T. and Kurtilla, T. (2000) 'Perceiving the environment in Finnish Lapland', *Body and Society* 6: 183–196.
Iveson, K. (2007) *Publics and the City*. Oxford: Wiley-Blackwell.
Jacobs, J. (2006) 'A geography of big things', *Cultural Geographies* 13: 1–27.
Jacobs, J. and Smith, S. (2008) 'Living room: rematerialising home', *Environment and Planning A* 40: 515–519.
Jazeel, T. (2007) 'Awkward geographies: spatializing academic responsibility, encountering Sri Lanka', *Singapore Journal of Tropical Geography* 28: 287–299.
Jazeel, T. and McFarlane, C. (2007) 'Responsible learning: cultures of knowledge production and the North–South divide', *Antipode* 39: 781–789.
Jazeel, T. and McFarlane, C. (2010) 'The limits of responsibility: a postcolonial politics of academic knowledge production', *Transactions of the Institute of British Geographers* 35: 109–124.
Jeffrey, C. and McFarlane, C. (2008) 'Guest editorial: performing cosmopolitanism', *Environment and Planning D: Society and Space* 26: 420–442.
Johnson, H. (2007) 'Communities of practice and international development', *Progress in Development Studies* 7: 277–290.
Johnson, H. and Wilson, G. (2009) *Learning for Development*. London: Zed.
Joyce, P. (2003) *The Rule of Freedom: Liberalism and the Modern City*. London: Verso.
Juris, J. (2008) *Networking Futures: The Movements Against Corporate Globalization*. Durham, NC: Duke University Press.
Kaldor, M. (2003) *Global Civil Society: An Answer to War*. Cambridge: Polity.
Kantor, P. and Savitch, H.V. (2005) 'How to study comparative urban development politics: a research note', *International Journal of Urban and Regional Research* 29: 135–151.
Keck, M. and Sikkink, K. (1998) *Activists Beyond Borders: Advocacy Networks in International Politics*. Ithaca, NY: Cornell University Press.
Keith, M. (2005) *After the Cosmopolitan*. London: Routledge.
King, A.D. (1976) *Colonial Urban Development: Culture, Social Power and Environment*. London and Boston: Routledge & Kegan Paul.
King, A.D. (1991a) *Global Cities: Post-Imperialism and the Internationalization of London*. New York: Russell Sage.
King, A.D. (1991b) *Urbanism, Colonialism, and the World-Economy: Cultural and Spatial Foundations of the World Urban System*. London: Routledge.
King, A.D. (2004) *Spaces of Global Cultures: Architecture, Urbanism, Identity*. London: Routledge.
King, K. and McGrath, S. (2004) *Knowledge for Development? Comparing British, Japanese, Swedish and World Bank Aid*. London: Zed Books.
Kofman, E. and Lebas, E. (1996) *Henri Lefebvre: Writing on Cities*. Oxford: Blackwell.
Kraftl, P. and Adey, P. (2008) 'Architecture/affect/inhabitation: geographies of being-in buildings', *Annals of the Association of American Geographers* 98: 213–231.

Kumbaya-Senkwe, B.M. and Lumambo, P. (2010) 'Subaltern speak in a postcolonial setting: diffusing and contesting donor-engendered knowledge in the water sector in Zambia', in P. Healey and R. Upton (eds), *Crossing Borders: International Exchange and Planning Practices*. London: Routledge, pp. 191–218.

Larner, W. (2003) 'Neoliberalism?', *Environment and Planning D: Society and Space* 21: 509–512.

Larner, W. and Le Heron, R. (2002) 'The spaces and subjects of a globalising economy: a situated exploration of method', *Environment and Planning D: Society and Space* 20: 753–774.

Larner, W. and Laurie, N. (2010) 'Travelling technocrats, embodied knowledges: globalising privatisation in telecoms and water', *Geoforum* 41: 218–226.

Latour, B. (1986) 'The power of association', in J. Law (ed.), *Power, Action and Belief: A New Sociology of Knowledge?* London: Routledge and Kegan Paul.

Latour, B. (1999) *Pandora's Hope: Essays on the Reality of Science Studies*. Cambridge, MA: Harvard University Press.

Latour, B. (2004) *Politics of Nature: How to Bring the Sciences into Democracy* (trans. C. Porter). Cambridge, MA: Harvard University Press.

Latour, B. (2005) *Reassembling the Social: An Introduction to Actor-Network Theory*. Oxford: Clarendon.

Lave, J. (1988) *Cognition in Practice: Mind, Mathematics and Culture in Everyday Life*. Cambridge: Cambridge University Press.

Law, J. (2000) 'Transitivities', *Environment and Planning D: Society and Space* 18: 133–148.

Leach, M. and Coones, I. (2007) 'Mobilising citizens: social movements and the politics of knowledge', *IDS Working Paper 276*, January 2007.

Leadbeater, C. (2000) *Living on Thin Air: The New Economy*. London: Penguin.

Leavitt, J. (1994) 'Planning in an age of rebellion: guidelines to activist research and applied planning', *Planning Theory* 10: 111–130.

Le Corbusier (2008 [1923]) *Towards a New Architecture*. BN Publishing.

Lees, L., Slater, T. and Wyly, E. (2008) *Gentrification*. New York: Routledge.

Lefebvre, H. (1971) *Everyday Life in the Modern World*. New York: Harper and Row.

Lefebvre, H. (1991) *Critique of Everyday Life*. London: Verso.

Lefebvre, H. (2004 [1970]) *The Urban Revolution*. Minneapolis: University of Minnesota Press.

Lefebvre, H. (2004) *Rhythmanalysis: Space, Time and Everyday Life* (trans. S. Elden). London: Continuum.

Le Gales, P. (2002) *European Cities: Social Conflicts and Governance*. Oxford: Oxford University Press.

Legg, S. (2007) *Delhi's Urban Governmentalities*. Oxford: Wiley-Blackwell.

Legg, S. (2009) 'Of scales, networks and assemblages: the League of Nations apparatus and the scalar sovereignty of the Government of India', *Transactions of the Institute of British Geographers* 34: 234–253.

Leith, A. (1864) *Report on the Sanitary State of the Island of Bombay*. Bombay: Education Society Press.

Leitner, H., Sheppart, E. and Sziatro, K.S. (2008) 'The spatialities of contentious politics', *Transactions of the Institute of British Geographers* 33: 157–172.

Ley, A. (2010) *Housing as Governance. Interfaces between Local Government and Civil Society Organisations in Cape Town, South Africa*. Berlin: LIT Verlag.

Leydesdorff, L. (2006) *The Knowledge-Based Economy: Modeled, Measured, Simulated*. Florida: Universal Publishers.

Li, T.M. (2007) 'Practices of assemblage and community forest management', *Economy and Society* 36: 263–293.

Li, T.M. (2008) 'Practices of assemblage and community forest management', *Economy and Society* 36: 263–293.

Lingis, A. (1996) *Sensation: Intelligibility in Sensibility*. New Jersey: Humanities Press.

Lomnitz, L. (1977) *Networks and Marginality: Life in a Mexican Shantytown*. New York: Academic Press.

London, C. (2002) *Bombay Gothic*. Mumbai: India Book House.

Lorrain, D. (2005) Urban capitalisms: European models in competition. *International Journal of Urban and Regional Research* 29: 231–267.

Lyons, M., Smut, C. and Stephens, A. (2001) 'Participation, empowerment and sustainability: (how) do the links work?' *Urban Studies* 38: 1233–1251.

Machlup, F. (1962) *Production and Distribution of Knowledge*. Princeton: Princeton University Press.

MacKinnon, D. (2008) 'Evolution, path-dependence, and economic geography', *Geography Compass* 2: 1449–1463.

MacLeod, G. (2002) 'From urban entrepreneurialism to a "Revanchist City"? On the spatial injustices of Glasgow's renaissance', *Antipode* 34: 602–624.

Marcuse, H. (1972) *Counter-Revolution and Revolt*. East Sussex: Beacon Press.

Marcuse, P. (2009) 'From critical urbanism to right to the city', *City* 13: 185–197.

Masser, I. and Williams, R.H. (1986) *Learning from Other Countries: The Cross-National Dimension in Urban Policy-Making*. Norwich: Geobooks/Elsevier.

Massey, D. (2005) *For Space*. London: Sage.

Massey, D. (2007) *World City*. Cambridge: Polity.

Massey, D. (forthcoming) 'A counterhegemonic relationality of place', in E. McCann, and K. Ward (eds), *Mobile Urbanism: Cities and Policy-making in the Global Age*. Minneapolis: University of Minnesota Press.

Mawdsley, E., Townsend, J.G., Porter, G. and Oakley, P. (2001) *Knowledge, Power and Development Agendas: NGOs North and South*. Oxford: INTRAC.

Maxwell, S. (1998) 'Comparisons, convergences and connections: development studies in North and South', in A. de Haan and S. Maxwell (eds), *Poverty and Social Exclusion in North and South*. Sussex: Institute of Development Studies Seminar Series, pp. 20–31.

Mbembe, A. (2004) 'Aesthetics of superfluity', *Public Culture* 16: 73–405.

Mbembe, A. and Nuttall, S. (2004) 'Writing the world from an African metropolis', *Public Culture* 16: 347–372.

McCann, E.J. (2008) 'Expertise, truth, and urban policy mobilities: global circuits of knowledge in the development of Vancouver, Canada's 'Four Pillar' drug strategy', *Environment and Planning A* 40: 885–904.

McCann, E.J. (2011) 'Urban policy mobilities and global circuits of knowledge: toward a research agenda', *Annals of the Association of American Geographers* 101: 107–130.

McCann, E.J. and Ward, K. (2010) 'Relationality/territoriality: toward a conceptualization of cities in the world', *Geoforum* 41: 175–184.

McCann, E.J. and Ward, K. (2011) *Mobile Urbanism: City Policymaking in the Global Age*. Minneapolis: University of Minnesota Press.

McDonough, T. (ed.) (2009) *The Situationists and the City*. London: Verso.

McEwan, C. (2003) 'Material geographies and postcolonialism', *Singapore Journal of Tropical Geography* 24: 340–355.

McFarlane, C. (2004) 'Geographical imaginations and spaces of political engagement: examples from the Indian Alliance', *Antipode* 36: 890–916.

McFarlane, C. (2006a) 'Knowledge, learning and development: a post-rationalist approach', *Progress in Development Studies* 6: 287–305.

McFarlane, C. (2006b) 'Transnational development networks: bringing development and postcolonial approaches into dialogue', *The Geographical Journal* 172: 35–49.

McFarlane, C. (2006c) 'Crossing borders: development, learning and the North-South divide', *Third World Quarterly* 27: 1413–1438.

McFarlane, C. (2008a) 'Governing the contaminated city: sanitation in colonial and postcolonial Bombay', *International Journal of Urban and Regional Research* 32: 415–435.

McFarlane, C. (2008b) 'Postcolonial Bombay: decline of a cosmopolitan city?' *Environment and Planning D: Society and Space* 26: 480–499.

McFarlane, C. (2008c) 'Sanitation in Mumbai's informal settlements: state, "slum" and infrastructure', *Environment and Planning A* 40: 88–107.

McFarlane, C. (2009a) 'Translocal assemblages: space, power and social movements', *Geoforum* 40: 561–567.

McFarlane, C. (2009b) 'Infrastructure, interruption and inequality: urban life in the global South', in S. Graham (ed.), *Disrupted Cities: When Infrastructure Fails*. London: Routledge, pp. 131–144.

McFarlane, C. (2010) 'The comparative city: knowledge, learning, urbanism', *International Journal of Urban and Regional Research* 34: 725–742.

McFarlane, C. (2011a) 'The city as assemblage: dwelling and urban space', *Environment and Planning D: Society and* Space (forthcoming).

McFarlane, C. (2011b) 'Assemblage and critical urbanism', *City* (forthcoming).

McGuirk, P. and Dowling, R. (2009) 'Neoliberal privatisation? Remapping the public and the private in Sydney's masterplanned residential estates', *Political Geography* 28: 174–185.

McLennan, G. (2004) 'Travelling with vehicular ideas: the case of the third way', *Economy and Society* 33: 484–499.

McNeill, D. (2009) *The Global Architect: Firms, Fame, and Urban Form*. New York and Abingdon: Routledge.

Mehta, L. (1999) 'Knowledge for development: World Development Report 1998/99', *Journal of Development Studies* 36: 151–161.

Mehta, L. (2001) 'The World Bank and its emerging knowledge empire', *Human Organization* 60: 89–96.

Melucci, A. (1989) *Nomads of the Present: Social Movements and Individual Needs in Contemporary Society* (eds Keane and Mier). Philadelphia, PA: Temple University Press.

Merrifield, A. (2002) *Metromarxism: a Marxist tale of the city*. London: Routledge.

Middleton, J. (2009) '"Stepping in time": walking, time, and space in the city', *Environment and Planning A* 41: 1943–1961.

Mitchell, D. (2003) *The Right to the City: Social Justice and the Fight for Public Space*. New York: Guilford Place.

Mitlin, D. (2008) *Urban Poor Funds: Development by the People for the People*. Poverty Reduction in Urban Areas Series: Working Paper 18. London: Institute for International Environment and Development.

Mohan, G. and Stokke, K. (2000) Participatory development and empowerment: the convergence around civil society and the dangers of localism. *Third World Quarterly* 21: 247–268.

Mohapatra, B.N. (2003) 'Civil society and governance: from the vantage point of the pavement dwellers of Mumbai', in R. Tandon and R. Mohanty (eds), *Does Civil Society Matter? Governance in Contemporary India*. London: Sage, pp. 285–314.

Mol, A.M. (2009) *The Logic of Care: Health and the Problem of Patient Choice*. London: Routledge.

Monbiot, G. (2004) 'This is what we are paid for', *Guardian*, 18 May. http://www.guardian.co.uk/politics/2004/may/18/foreignpolicy.india (accessed November 2010).

Morton, P. (2000) *Hybrid Modernities. Architecture and Representation at the 1931 Colonial Exposition, Paris*. Cambridge, MA: MIT Press.

Moser, C. (1996) *Confronting Crisis: A Comparative Study of Household Responses to Poverty and Vulnerability in Four Poor Urban Communities*. Washington, DC: World Bank.

Mouffe, C. (2000) *The Democratic Paradox*. London: Verso.

Mufti, A. (2005) 'Global comparativism', *Critical Inquiry* 31: 427–489.

Munt, S.R. (2001) 'The Lesbian Flâneur', in I. Borden, J. Kerr, J. Rendell and A. Pivaro (eds), *The Unknown City: Contesting Architecture and Social Space*. Cambridge, MA: MIT Press, pp. 246–261.

Naidu, N.C. and Ninan, S. (2000) *Plain Speaking*. New Delhi: Viking.

Napolitano, V. and Pratten, D. (2007) 'Michel De Certeau: ethnography and the challenge of plurality', *Social Anthropology* 15: 1–12.

Nasr, J. and Volait, M. (eds) (2003) *Urbanism: Imported or Exported?* Oxford: Blackwell.

Negri, A. (2006) *The Porcelain Workshop: For a New Grammar of Politics*. Los Angeles: Semiotext(e).

Neuwirth, R. (2006) *Shadow Cities: A Billion Squatters, A New Urban World*. London: Routledge.

Nicholls, W.J. (2008) 'The urban question revisited: the importance of cities for social movements', *International Journal of Urban and Regional Research* 32: 841–859.

Nijman, J. (2007a) 'Introduction: comparative urbanism', *Urban Geography* 28: 1–6.

Nijman, J. (2007b) 'Place-particularity and "deep analogies": a comparative essay on Miami's rise as a world city', *Urban Geography* 28: 92–107.

Noble, G. (2004) 'Accumulating being', *International Journal of Cultural Studies* 7: 233–256.

Nonaka, I., Toyama, R. and Konno, N. (2000) 'SECI, Ba and leadership: a unified model of dynamic knowledge creation', *Long Range Planning* 33: 5–34.

Obrador-Pons, P. (2003) 'Being-on-holiday: tourist dwelling, bodies and place', *Tourist Studies* 3: 47–66.

Olds, K. (2001) *Globalization and Urban Change: Capital, Culture, and Pacific Rim Mega-Projects*. Oxford: Oxford University Press.

Olds, K. and Thrift, N. (2005) 'Assembling the "global schoolhouse" in Pacific Asia', in P. Daniels, K.C. Ho and T. Hutton (eds), *Service Industries, Cities and Development Trajectories in the Asia-Pacific*. London: Routledge, pp. 201–216.

Oliviera, N.D.S. (1996) 'Favelas and ghettos: race and class in Rio de Janeiro and New York City', *Latin American Perspectives* 90: 71–89.

Olssen, M. (2008) 'Understanding the mechanisms of neoliberal control: lifelong learning, flexibility and knowledge capitalism', in A. Fejes and K. Nicoll (eds), *Foucault and Lifelong Learning: Governing the Subject*. London: Routledge.

Ong, A. (1999) *Flexible Citizenship: The Cultural Logics of Transnationality*. Durham and London: Duke University Press.

Ong, A. (2007) 'Neoliberalism as a mobile technology', *Transactions of the Institute for British Geographers* 32: 3–8.

Ong, A. and Collier, S.J. (eds) (2005) *Global Assemblages: Technology, Politics and Ethics as Anthropological Problems*. Oxford: Blackwell.

Otter, C. (2004) Cleansing and clarifying: technology and perception in nineteenth-century London. *Journal of British Studies* 43: 40–64.

Patel, S. (no date) 'How can poor people benefit from research results?' Unpublished paper. Bombay: Society for Promotion of Area Resource Centres (SPARC).

Patel, S. (1997) *From the Slums of Bombay to the Housing Estates of Britain: A Look into Community Involvement in the Process of Community Regeneration in Britain, Sharing Ideas and Practice from Working with Slum Communities in India*. London: Centre for Innovation in Voluntary Action/Oxfam.

Patel, S., Burra, S. and D'Cruz, C. (2001) 'Slum/Shack Dwellers International (SDI) – foundations to treetops', *Environment and Urbanization* 13: 45–59.

Patel, S. and Mitlin, D. (2001) 'The work of SPARC, the National Slum Dwellers Federation and Mahila Milan', *International Institute for Environment and Development (IIED): Poverty Reduction in Urban Areas Series, Working Chapter 5*. London: IIED.

Patel, S. and Mitlin, D. (2002) 'Sharing experiences and changing lives', *Community Development Journal* 37: 125–136.

Patnaik, U. (1999) *The Long Transition: Essays on Political Economy*. New Delhi: Tulika.

Peattie, L. (1994) 'Communities and interests in advocacy planning', *Journal of the Amerian Planning Association* 60: 151–153.

Peck, J. (2003) 'Geography and public policy: mapping the penal state', *Progress in Human Geography* 27: 222–232.

Peck, J. (2005) 'Struggling with the creative class', *International Journal of Urban and Regional Research* 29: 740–770.

Peck, J. (2006) 'Liberating the city: between New York and New Orleans', *Urban Geography* 27: 681–713.

Peck, J. (forthcoming) 'Recreative city: Amsterdam, vehicular idea, and the adaptive spaces of creativity policy'. Unpublished manuscript.

Peck, J. and Theodore, N. (2001) 'Exporting workfare/importing welfare-to-work: exploring the politics of Third Way policy transfer', *Political Geography* 20: 427–460.

Peck, J. and Theodore, N. (2008) 'Carceral Chicago: making the ex-offender employability crisis', *International Journal of Urban and Regional Research*, online.

Peck, J. and Theodore, N. (2010) 'Mobilizing policy: models, methods and mutations', *Geoforum* 41: 169–174.
Perera, N. (2005) 'Importing urban problems: the impact of the introduction of the Housing Ordinance in Colombo', *Arab World Geographers* 7: 1–2.
Phillips, J. (2006) 'Agencement/assemblage', *Theory, Culture and Society* 23: 108–109.
Philo, C. and Parr, H. (2000) 'Editorial. Institutional geographies: introductory remarks', *Geoforum* 31: 513–521.
Pieterse, E. (2008) *City Futures: Confronting the Crisis of Urban Development*. London: Zed Books.
Pieterse, J.N. (1998) 'My paradigm or yours? Alternative development, post-development, reflexive development', *Development and Change* 29: 343–373.
Pieterse, J.N. (2001) *Development Theory: Deconstructions/Reconstructions*. London: Sage.
Pile, S. (2001) 'The un(known) city … or, an urban geography of what lies buried below the surface', in I. Borden, J. Kerr, J. Rendell and A. Pivaro (eds), *The Unknown City: Contesting Architecture and Social Space*. Cambridge, MA: MIT Press, pp. 262–279.
Polanyi, M. (1966) *The Tacit Dimension* (1983 reprint). New York: Doubleday.
Polanyi, M. (1969) *Knowing and Being* (edited with an introduction by Marjorie Grene). Chicago: University of Chicago Press.
Popkewitz, T.S., Olsson, U. and Petersson, K. (2006) 'The learning society, the unfinished cosmopolitan, and governing education, public health and crime prevention at the beginning of the twenty-first century', *Educational Philosophy and Theory* 38: 431–449.
Prakash, G. (2006) 'The idea of Bombay', *The American Scholar* 75: 88–99.
Prakash, G. (2010) *Mumbai Fables*. Princeton: Princeton University Press.
Prince, R. (2010) 'Policy transfer as policy assemblage: making policy for the creative industries in New Zealand', *Environment and Planning A* 42: 169–186.
Provoost, M. (2006) 'New towns on the Cold War frontier', Eurozine, www.eurozine.com/articles/2006-06-28-provoost-en.html, accessed February 2009.
Pyla, P. (2008) 'Back to the future: Doxiadis' plans for Baghdad', *Journal of Planning History* 7: 3–19.
Rabinow, P. (1989) *French Modern: Norms and Forms of the Social Environment*. Chicago: University of Chicago Press.
Rabinow, P. (2003) *Anthropos Today*. Princeton, NJ: Princeton University Press.
Rajchman, J. (1998) *Constructions*. Cambridge, MA: MIT Press.
Ramsamy, E. (2006) *The World Bank and Urban Development*. London: Routledge.
Rendell, J. (2006) *Art and Architecture: A Place Between*. London: I.B. Tauris.
Riles, A. (2001) *The Network Inside Out*. Ann Arbor: University of Michigan Press.
Riles, A. (ed.) (2006) *Documents: Artifacts of Modern Knowledge*. Ann Arbor: University of Michigan Press.
Risse-Kappen, T. (ed.) (1995) *Bringing Transnational Relations Back-In: Non-State Actors, Domestic Structures and International Institutions*. Cambridge: Cambridge University Press.
Robinson, J. (2002) 'Global and world cities: a view from off the map', *International Journal of Urban and Regional Change* 26: 513–554.
Robinson, J. (2006) *Ordinary Cities: Between Modernity and Development*, London: Routledge.

Robinson, P. (1998) 'Beyond workfare: active labour-market policies', in A. De Haan and S. Maxwell (eds), *Poverty and Social Exclusion in North and South*. Sussex: Institute of Development Studies, Seminar Series 29, pp. 86–93.

Rose, R. (1991) 'What is lesson drawing?', *Journal of Public Policy* 11: 3–30.

Rose, R. (1993) *Lesson-Drawing in Public Policy: A Guide to Learning Across Time and Space*. New York: Chatham House Publishers Inc.

Routledge, P. and Cumbers, A. (2009) *The Entangled Geographies of Global Justice Networks*. Manchester: Manchester University Press.

Roy, A. (2003) 'Paradigms of propertied citizenship: transnational techniques of analysis', *Urban Affairs Review* 38: 463–490.

Roy, A. (2005) 'Urban informality: towards an epistemology of planning', *Journal of the American Planning Association* 71: 147–158.

Roy, A. (2009a) 'Civic governmentality: the politics of inclusion in Beirut and Mumbai', *Antipode* 41: 159–179.

Roy, A. (2009b) 'Why India cannot plan its cities: informality, insurgence and the idiom of urbanization', *Planning Theory* 8: 76–87.

Ruddick, S. (1996) *Young and Homeless in Hollywood: Mapping Social Identities*. New York: Routledge.

Said, E. (1984) *The World, the Text and the Critic*. London: Faber and Faber.

Said, E. (1993) *Culture and Imperialism*. London: Chatto and Windus.

Sainath, P. (2004) 'Chandrababu: image and reality', *The Hindu*, 5 July. http://www.thehindu.com/2004/07/05/stories/2004070503400800.htm (accessed November 2010).

Sandercock, L. (1998) *Towards Cosmopolis: Planning for Multicultural Cities*. London: John Wiley.

Sandercock, L. (2004) *Cosmopolis II*. London: Continuum.

Sassen, S. (2003) 'Globalization or denationalisation?', *Review of International Political Economy* 10: 1–22.

Sassen, S. (2007) *Territory, Authority, Rights: From Medieval to Global Assemblages*. Princeton: Princeton University Press.

Saunier, P.Y. (1999) 'Changing the city: urban international information and the Lyon municipality, 1900–1940', *Planning Perspectives* 14: 19–48.

Saunier, P.Y. (2001) 'Sketches from the Urban Internationale, 1910–50: voluntary associations, international institutions and US philanthropic foundations', *International Journal of Urban and Regional Research* 25: 380–403.

Saunier, P.Y. (2002) 'Taking up the bet on connections: a municipal contribution', *Contemporary European History* 11: 507–527.

Savage, M. (2009) 'Contemporary sociology and the challenge of descriptive assemblage', *European Journal of Social Theory* 12: 155–174.

Schatzki, T. (2001) *The Practice Turn in Contemporary Theory*. London: Routledge.

Schneider, J. and Susser, I. (eds) (2003) *Wounded Cities: Deconstruction and Reconstruction in a Globalized World*. New York: Berg.

Schumpeter, J. (1934) *Theory of Economic Development*. London: Harvard University Press.

Scott, A.J. (2006) 'A perspective of economic geography', in S. Bagchi-Sen and H. Lawton Smith (eds) *Economic Geography: Past, Present and Future*. Oxford: Routledge, pp. 56–80.

Scott, J.C. (1985) *Weapons of the Weak: Everyday Forms of Peasant Resistance.* New Haven, CT: Yale University Press.

Scott, J.C. (1990) *Domination and the Arts of Resistance: Hidden Transcripts.* New Haven, CT: Yale University Press.

Seamon, D. (ed.) (1993) *Dwelling, Seeing, Building: Toward a Phenomenological Ecology.* Albany: State University of New York Press.

Seamon, D. (1998) 'Concretizing Heidegger's notion of dwelling: the contributions of Thomas Thiis-Evensen and Christopher Alexander', *Subject*, www.tu-cottbus/de.BTU/Fak2/TheoArch/Wolke/eng/Subjects/982/Seamon (accessed September 2009).

Seamon, D. (2000) 'Phenomenology, Place, Environment and Architecture: A Review'. http://www.arch.ksu.edu/seamon/Seamon_reviewEAP.htm (accessed December 2010).

Sennett, R. (2008) *The Craftsman.* London: Allen Lane.

Serres, M. (1974) *La Traduction, Hermes III.* Paris: Les Editions de Minuit.

Serres, M. and Latour, B. (1995) *Conversations in Science, Culture and Time.* Ann Arbor, MI: University of Michigan Press.

Sharan, A. (2006) 'In the city, out of place: environment and modernity, Delhi 1860s to 1960s', *Economic and Political Weekly* November 25: 4905–4911.

Sharma, K. (2001) Housing Mumbai's millions. *The Hindu.* Online edition: 1 February.

Shove, E., Watson, M., Hand, M. and Ingram, J. (2007) *The Design of Everyday Life.* Oxford: Berg.

Silva, R.T. (2000) 'The connectivity of infrastructure networks and the urban space of São Paulo', *International Journal of Urban and Regional Research* 24: 139–164.

Simone, A. (2004a) 'People as infrastructure: intersecting fragments in Johannesburg', *Public Culture* 16: 407–429.

Simone, A. (2004b) *For the City Yet to Come: Changing African Life in Four Cities.* Durham, NC: Duke University Press.

Simone, A. (2008a) 'Emergency democracy and the "governing composite"', *Social Text* 26: 13–33.

Simone, A. (2008b) 'The politics of the possible: making urban life in Phnom Penh', *Singapore Journal of Tropical Geography* 29: 186–204.

Simons, M. (2006) 'Learning as investment: notes on governmentality and biopolitics', *Educational Philosophy and Theory* 38: 523–539.

Sinclair, I. (1997) *Lights Out for the Territory: Nine Excursions in the Secret History of London.* London: Granta.

Sintomer, Y., Herzberg, C. and Rocke, A. (2008) 'Participatory budgeting in Europe: potentials and challenges', *International Journal of Urban and Regional Research* 32: 164–178.

Smith, M.E. (2009) 'Editorial. Just how comparative is comparative urban geography? A perspective from Archaeology', *Urban Geography* 30: 113–117.

Smith, M.P. (2001) *Transnational Urbanism: Locating Globalization.* Malden, MA: Blackwell.

Smith, N. (1996) *The New Urban Frontier: Gentrification and the Revanchist City.* London: Routledge.

Smith, N. (1998) 'Giuliani time: the revanchist 1990s', *Social Text* 57: 1–20.

Smith, N. (2002) 'New globalism, new urbanism: gentrification as global urban strategy', *Antipode* 34: 434–457.
SPARC (1988) *Beyond the Beaten Track: Resettlement Initiatives of People who Live Along the Railway Tracks in Bombay*. Mumbai: Society for the Promotion of Area Resource Centres.
Spivak, G.C. (1993) *Outside in the Teaching Machine*. London: Routledge.
Spivak, G.C. (2003) *Death of a Discipline*. New York: Columbia University Press.
Star, S.L. (1999) 'The ethnography of infrastructure', *American Behavioural Scientist* 43: 377–391.
Stead, D., De Jong, M. and Reinholde, I. (2010) 'West–East policy transfer in Europe: the case of urban transport policy', in P. Healey and R. Upton (eds), *Crossing Borders: International Exchange and Planning Practices*. London: Routledge, pp. 173–190.
Stengers, I. (1997) *Power and Invention: Situating Science*. Minneapolis: University of Minnesota Press.
Stone, D. and Denham, A. (eds) (2004) *Think Tank Traditions: Policy Research and the Politics of Ideas*. Manchester: University of Manchester Press.
Stone, D. and Maxwell, S. (eds) (2004) *Global Knowledge Networks and International Development: Bridges Across Boundaries*. London: Routledge.
Storper, M. and Scott, A.J. (2009) 'Rethinking human capital, creativity and urban growth', *Journal of Economic Geography* 9: 147–167.
Strobel, R.W. (2003) 'From "cosmopolitan fantasies" to "nationalist traditions": socialist realism in East Berlin', in J. Nasr and M. Volait (eds), *Urbanism: Imported or Exported?* Oxford: Blackwell, pp. 128–154.
Sutcliffe, A. (1981) *Towards the Planned City, Germany, Britain, the United States and France, 1780–1914*. Oxford: Blackwell.
Swanson, K. (2007) 'Revanchist urbanism heads south: the regulation of indigenous beggars and street vendors in Ecuador', *Antipode* 39: 708–728.
Swyngedouw, E. (2006) 'Circulations and metabolisms: (hybrid) natures and (cyborg) cities', *Science as Culture* 15: 105–121.
Tampio, N. (2009) 'Assemblages and the multitude: Deleuze, Hardt, Negri, and the Postmodern Left', *European Journal of Political Theory* 8: 383–400.
Theodosis, L. (2008) ' "Containing" Baghdad: Constanitnos Doxiadis' Program for a Developing Nation', in *Revista de crítica arquitectónica*. Ciudad del Espejismo: Bagdad, de Wright a Venturi, pp. 167–172.
Thrift, N. (2007) *Non-Representational Theory: Space, Politics, Affect*. London: Routledge.
Tomlinson, R. (2002) 'International best practice, enabling frameworks, and the policy process: a South African case study', *International Journal of Urban and Regional Research* 26: 377–388.
Turner, J.F.C. (1972) 'Housing as a verb', in J.F.C. Turner and R. Fichter (eds), *Freedom to Build*. New York: Macmillan, pp. 148–175.
UNDP (2001) *Human Development Report, 2001*. New York: UNDP.
UNDP (2003) *Partnership for Local Capacity Development: Building on the Experiences of City-to-City Cooperation*. New York: UNDP.
Van Loon, J. (2006) 'Network', *Theory, Culture and Society* 23: 307–314.
Vasudevan, A., McFarlane, C. and Jeffrey, A. (2008) 'Spaces of enclosure', *Geoforum* 39: 1641–1646.

Venn, C. (2006a) 'A note on assemblage', *Theory, Culture and Society* 23: 107–108.
Venn, C. (2006b) 'The city as assemblage: diasporic cultures, postmodern spaces, and biopolitics', in H. Berking, S. Frank, L. Frers, L.M. Low, S. Steets and S. Stoetzer (eds), *Negotiating Urban Conflicts: Interaction, Space and Control*. London: Transaction.
Vincentian Missionaries Social Development Incorporated (VMSDFI) (2001) 'Meet the Philippines Homeless People's Federation', *Environment and Urbanization* 13: 73–84.
Wacquant, L. (2007) *Urban Outcasts: A Comparative Sociology of Advanced Marginality*. Cambridge: Polity.
Wacquant, L. (2008) 'The militarization of urban marginality: lessons from the Brazilian metropolis', *International Political Sociology* 2: 56–74.
Wang, B. (2010) 'Cities in transition: episodes of spatial planning in modern China', in P. Healey and R. Upton (eds), *Crossing Borders: International Exchange and Planning Practices*. London: Routledge, pp. 95–116.
Ward, K. (2006) 'Policies in motion, urban management and state restructuring: the trans-local expansion of Business Improvement Districts', *International Journal of Urban and Regional Research* 30: 54–70.
Ward, K. (2007) 'Business Improvement Districts: policy origins, mobile policies and urban liveability', *Geography Compass* 1: 657–672.
Ward, K. (2008) 'Commentary: toward a comparative (re)turn in urban studies? Some reflections', *Urban Geography* 29: 1–6.
Ward, K. (2009) 'Towards a relational comparative approach to the study of cities', *Progress in Human Geography* 34: 471–487.
Watson, S. (1988) *Accommodating Inequality*. Sydney: Allen and Unwin.
Wenger, E. (1998) *Communities of Practice: Learning, Meaning, and Identity*. Cambridge: Cambridge University Press.
Whatmore, S. (2006) 'Materialist returns: practising cultural geography in and for a more-than-human world', *Cultural Geographies* 13: 600–609.
Wigley, M. (2001) 'Network fever', *Grey Room* 4: 82–122.
Williams, G. (2004) 'Evaluating participatory development: tyranny, power and (re) politicisation', *Third World Quarterly* 25: 557–579.
Winkler, J. (2002) 'Working on the experience of passing environments: on commented walks', in J. Winkler (ed.), *Space, Sound and Time: A Choice of Articles in Soundscape Studies and Aesthetics of Environment 1990–2003*. http://www.humgeo.unibas.ch/homepages/winkler.htm (accessed December 2010).
World Bank (1999) *World Development Report 1998/9: Knowledge for Development*. Oxford: World Bank/Oxford University Press.
Wright, G. (1991) *The Politics of Design in French Colonial Urbanism*. Chicago: University of Chicago Press.
Wunderlich, F.M. (2008) 'Walking and rhythmicity: sensing urban space', *Journal of Urban Design* 13: 125–139.
Yaeger, P. (2007) 'Introduction: dreaming of infrastructure', *Publications of the Modern Language Association of America (PMLA)* 122: 9–26.
Yiftachel, O. (2009) 'Critical theory and "grey space": mobilization of the colonized', *City* 13: 240–256.
Young, R.J.C. (2001) *Postcolonialism. An Historical Introduction*. Oxford: Blackwell.

Index

Abers, R. 99, 100, 104
ACHR (Asian Coalition for Housing Rights) 70–1, 75, 76, 183
Adey, P. 21
Adorno, T. 157
advocacy networks 64
Africa 37
 see also under various country names, e.g.: Botswana; Egypt; Ghana; Kenya; Malawi; Mozambique; Namibia; Nigeria; South Africa; Sudan; Uganda; Zambia; Zimbabwe
agency 25, 26, 27, 69, 81, 149, 155, 166
 collective 19
 critical learning and 160–4
 materials 63
 reconfiguring of 167, 173, 179
 structure and 90
Akrich, M. 78
alert reverie 20–1
Allen, J. 24, 68, 119–21, 127
Allende, Salvador 147
AlSayyad, N. 39
Amelang, J. 168
Amin, A. 3, 4, 5, 17–18, 20, 22, 93, 94, 119

Amritsar (Golden Temple) 42
Amsterdam 138
Anderson, B. 24
Anderson, E. 47
Andhra Pradesh 139
Anjaria, J. S. 41
Annan, Kofi 74
Annandi 73
ANT (actor-network theory) 24, 26, 78
Anzorena, J. 71–2
Appadurai, A. 68, 69, 74, 76, 84, 87, 170, 171, 180
Arputham, J. 75–6, 82, 161–2
Arthur Road jail 46
assassins 46
assemblage 1, 6, 9–31
 critical learning and 156, 160, 164–7
 distributed 3
 housing 69–74
 ideological 115–52
 learning 37, 41, 53, 60, 64
 legal 57–9
 see also urban learning assemblages
Atkinson, R. 94
Auckland 138
Austin, TX (US) 138, 150
Australia 118

Learning the City: Knowledge and Translocal Assemblage, First Edition. Colin McFarlane.
© 2011 Colin McFarlane. Published 2011 by Blackwell Publishing Ltd.

Badiou, A. 165
Baghdad 12, 116, 137, 146, 152
    Sadr City 129, 130
Baiocchi, G. 99, 100, 101
Bandra 49, 57
Banerjee-Guha, S. 88
Bangalore 75, 112, 140
Bangalore Slum Dwellers
    Federation 81–2
Barad, K. 163, 164
Barry, A. 26, 67
Barthe, Y. 96
Batliwala, S. 79
Bayat, A. 48
BDMM (Bangalore District
    Mahila Milan) 75
being-in-the-world 21
Beirut 137
Bender, T. 24
Benjamin, W. 41–2, 48, 51, 169
Bennett, J. 27, 35, 161, 163, 164
Berlin 48, 183
    see also East Berlin
Bharat Nagar 57, 58, 59
Bihar 44
Bingham, N. 78, 85
Birmingham 108, 166
BJP–Shiv Sena alliance 88
Blair, Tony 139
Blume, T. 108
Bollywood 46
Bolnick, J. 67
Bombay 12, 49, 116, 150
    sanitation 12, 13, 122–4, 125,
        126, 127
    see also Mumbai
Bombay First (NGO) 151
Borden, I. 2, 33, 37, 39–40, 51, 55
Botswana 110
Bourdieu, P. 22, 100
Bradford 106
Bratton, W. 12, 117, 141–2, 143
Brazil 10, 11, 85, 113, 143
    campaigning social movements 65
    see also Curitiba; Porto Alegre; Rio;
        São Paulo
Brenner, N. 153, 157, 168

Bridge, G. 47, 50, 51, 95
Briggs, J. 94
Bristol 108
Britain see Birmingham; Bradford;
    Bristol; Glasgow; Liverpool;
    Manchester; Plymouth;
    Southampton
budgeting see participatory budgeting
Bullock, M. 48
Bunnell, T. 4, 5, 117, 139, 141,
    148–9, 167
Bush, G. W. 138
bus-art production 42
Byculla 68, 81

Cabannes, Y. 106
Cairo 122
Caldeira, T. 102, 143
California 140
Callon, M. 5, 11, 15, 92, 96–8,
    104, 178
Calvino, Italo 174, 181
Campbell, T. 4
Canada 118
Cape Town 66, 68, 82
    Ekuphumleni 70
capitalism 28
Caribbean 122
Carnegie Trust 121
Castells, M. 48, 138
CBOs (community-based
    organizations) 66, 75, 83
census statistics 84–5
Chadwick, E. 123, 124
Chakrabarty, D. 125, 126
Chatterjee, P. 41
Chattopadhyay, S. 32, 42, 61, 122
Chicago School 169
Chile 147
China 133, 144, 161
    see also Shanghai
Chitekwe, B. 82–3
Christchurch (NZ) 106
Cities Alliance 62, 87, 106, 112, 183
citizenship 29, 94, 160
    insurgent 80
Clarke, N. 121, 149–50

classicism 131
Clifford, J. 29n, 184
Clinton, Bill 139
Cochrane, A. 24
codified knowledge 3, 19, 22, 56, 69,
    79, 82, 83
  instructions sedimented in 133
  process of creating 80
  translation of 59
Coe, N. M. 4, 141
Cohendet, P. 3, 4, 5, 17–18
Cold War 12, 63, 116, 130, 178
  US urban policy 128
Collier, S. J. 29–30
Colombo 122, 127
communicative action 64
communities of practice 4
comparative learning 18, 116,
    122–8, 180
  colonial 152
  postcolonial 155
competitive advantage 4
computers 79, 112
Connell, R. 169, 172
Conybeare, H. 123–7
Cooke, B. 94
Coones, I. 64–5
coordination 9, 16, 17, 19–20, 23, 25,
    30, 31, 54, 81, 97, 176
  importance of 10, 63
  lack of 77
  learning and 40, 59, 90–1, 98, 103
  multiple domains 80
Cornwall, A. 94
cosmopolitanism 22, 168, 170, 171,
    172, 180
  compositional 13, 155, 167, 173, 179
  existential 165
  inclusive 166
  progressive 164, 165–6
  subaltern street 44
  translocal 173, 179
Cost of Living Movement 99
Couto, C. G. 103
Coventry 11
craft traditions 43
Crang, M. 49

Crawford, A. 123
Cullen, J. 38
Cumbers, A. 4, 67
Curitiba 99
Cyberjaya 141

Dakar 71
Das, D. 5, 117, 139, 148–9
dasein 21
data-urbanism 78–85
De Certeau, M. 39, 50, 51, 54, 55,
    56, 60
De Haan, A. 110–11
De Landa, M. 24, 25, 27, 30, 156
De Soto, H. 86
Debord, G. 48, 51–2, 105
Deleuze, G. 24, 25, 27, 38, 155,
    158, 165
Delhi 44, 123, 126, 127, 134
Democratic Labour Party (Brazil) 99
dérive 51–2
DFID (UK Department for
    International Development) 87
Dialogue on Shelter (NGO) 77
divisions of labour 52
Dolowitz, D. 118
Dossal, M. 123, 124, 126
Dovey, K. 27, 38, 39, 52, 156–7
Dowling, R. 24, 30, 144, 147, 156
Doxiadis, C. 12, 116, 128–30, 133,
    148, 152
Dubai 140
Dublin 138
Durban 62
dwelling 16, 17, 25, 30, 31, 51, 56
  assemblages of 36, 48
  cosmopolitan 22
  crucial process of 33
  everyday 9, 55, 57
  incremental 37
  perception and 21–3
  plight of 2
  possibilities of/for 48, 53, 54, 55,
    57, 59
  see also housing; housing
    construction; learning-through-
    dwelling; urban dwelling

East Berlin 12, 116, 128, 130, 146, 152
  Frankfurter Tor 131
  Karl Marx Allee 131
East Germany *see* GDR
Easterling, K. 141
Ecuador 141, 142
  see also Guayaquil
EDP (computer firm) 79
education of attention 16, 21, 23, 31, 37, 60, 97
  dwelling as 9, 47, 49, 175–6
  walking as 50
Edwards, M. 109, 171
Egypt 112
Ekistics 128, 129, 146, 148
Elden, S. 21
Ellerman, D. 136
epistemic communities 119, 172
epistemic problem-spaces 11, 115, 145
epistemology 5, 16, 22, 135, 168, 169, 170
  displacement 9, 18
  inclusions and exclusions of 6
  Orientalist 122
  regional planning 147
  sociomaterial 9, 23, 31, 175
Epworth 77
Erundina, L. 103
ethico-politics 13, 16, 22, 155, 164, 168, 169–72, 173, 180
experiential expertise 65
explicit knowledge *see* codified knowledge
exteriority 25, 54
Eyerman, R. 63

Farías, I. 24, 156, 160
Featherstone, D. 2, 67, 68, 164
Federation of Tenants Association (Mumbai) 9, 57, 59, 102, 175, 176, 182
Fejes, A. 6, 141
feminist planning 95
Ferguson, J. 56, 65, 94, 120, 171
Fischer, G. 19–20
flâneurs 51, 169
Florida, R. 4, 5, 12, 28, 117, 138, 139, 140, 147

Ford Foundation 87, 121
Forrester, J. 95
Foucault, M. 5, 27, 84–5, 111n
France 121, 143
  see also Paris
Frankfurt School 157
Freire, P. 6, 103
French colonies 125n
Friedmann, J. 118, 147
FSIs (Floor Space Indices) 59, 88
Fuller, Buckminster 128

Gandy, M. 24, 28, 122
Gates Foundation 74, 87
Gauteng 71
Gaventa, J. 109–10, 114, 178
GDR (German Democratic Republic) 12, 131, 133
  see also East Berlin
Geddes, P. 121, 122
gentrification 138, 140, 141, 142, 153, 156, 162, 182
  growing trend towards 66
Germany 10, 106
  see also Berlin; GDR
Gertler, M. S. 4
Ghana 128
Gherardi, S. 3, 5, 17
Gibson, J. 16, 21
Gill, L. 40
Giuliani, Rudolph 141
Glaeser, E. L. 4
Glasgow 108, 124–5
global assemblage 29–30
Global Development Network 135
globalization 74, 137, 169
  anti-neoliberal movement 105
GLTN (Global Land Tool Network) 76, 77
Gobabis Municipality 113
governmentality 84, 85, 105, 120, 148
  ethnic 144
Grabher, G. 4, 20
Graham, S. 18, 39, 119, 137, 174
Green Bay, WI 138
Gregory, D. 18, 122
Groundswell 11, 93, 107, 108–9
  SPARC and 112, 114, 177, 178

Guattari, F. 27, 38
Guayaquil 116, 141, 142
Gulger, J. 168
Gupta, A. 120

Haas, P. 119
Habermas, J. 95, 98
Hansen, T. B. 32, 46, 47, 54, 55–6, 60–1, 128, 175
Harare 82, 116, 146
Haraway, D. 18
Hardt, M. 157–8
Harris, A. 122
Harrison, P. 21, 24
Harvard-MIT Center for Urban Studies 129
Harvey, D. 8, 90, 93, 134, 137, 140, 141, 161, 162, 165, 166, 173, 182, 184
Hayek, F. A. von 5
Healey, P. 49, 117
Heidegger, M. 2, 21–2, 36, 38, 159
Henselmann, H. 131
Heritage 134
Hetherington, K. 64
Hewlett, T. G. 123
Heynen, N. 161
Hinchliffe, S. 22, 53, 166
Hindu fundamentalists 88
HITECH (Hyderabad Information Technology Engineering Consultancy) 139
HIV/Aids 135
Hoffman, D. 144
Holland 143
Hollands, R. 5, 28–9, 138, 140, 143
Holloway, J. 78, 83
Holston, J. 80, 102
Homeless International (UK) 11, 107
homelessness 74, 108, 109, 177
  see also Groundswell; SAHPF
Horkheimer, M. 157
Hosagrahar, J. 123
housing 21, 32, 33, 48, 81, 84
  assistance to families 101
  basis for reimagining options 90, 177
  collective 68
  conditions 76, 80

  cooperative societies 58
  destruction/demolition of 40, 66, 162
  free 88
  gentrified 142
  government fund for subsidies 89
  incremental 36, 37, 39
  informal 55, 59, 86, 130, 162, 176
  low-income 36, 73, 136
  middle-class activists 57
  negotiations for 85
  public sector 101, 136
  substantial improvements in 103
  see also ACHR; Maharashtra Housing Board; MASHAL;
housing construction 9, 130, 148
  experiential knowledge of the poor for 71
  see also model house construction
Housing Ordinance (Colombo 1915) 122
Huchzermeyer, M. 76, 77, 83, 84, 85–6, 89, 168
Hurricane Katrina (US, 2002) 137
Hutchins, E. 5, 18, 19
Hyderabad 75, 139

Ibhert, O. 20
ICTs (information and communication technologies) 135
ideology
  and explanation 145–51
  and postwar urban planning 128–34
IDS (Institute of Development Studies) 110
immersion 21, 31, 35, 40, 56
  embodied 43
  experiential 32
  haptic 16, 47, 49
  practical 39
incremental learning 9, 38, 46, 59, 60, 69, 176
  reciprocal systems emerge through 40
  tactical learning can arise from 55
  urban space emerges through 39

India 10, 108, 109, 110, 112, 122
  high-tech service industries 144
  see also Andhra Pradesh; Bangalore;
    Delhi; Kolkata; Mumbai;
    Hyderabad; Pune; Uttar Pradesh
Indian Alliance 11, 36–7, 65, 67, 69,
    74, 75, 83, 87
  substantial gains for the poor 90
  WTI *(We, the Invisible)* 78, 79–81,
    82, 85
  see also SPARC
Indymedia 64
Infocity 4
infrapower 55, 56–7, 59, 60, 128, 175
Ingold, T. 1–2, 5, 15, 16, 21, 22, 48,
    49, 50, 53
intensity-openness-quality schema 98,
    103, 109, 112, 113
Inter-American Development
    Bank 106
Internacao 62
International Institute for Environment
    and Development 112
International Journal of Drug Policy 148
International Labour Office 121
Internet 4, 12, 17, 64, 116, 134, 135
  facilitating translocal urban learning
    forums 112
Iraq 128
  see also Baghdad

Jacobs, J. 21, 36, 148
Jamison, A. 63
Japan 107
Jazeel, T. 169, 171
Jeffrey, C. 164
Jennings, M. W. 48
Johannesburg 151, 169
Johnson, H. 77, 80
Joyce, P. 20
Juris, J. 64, 65, 112

Kaaba 42
Kabul 137
Kaldor, M. 63–4
Kanpur 75
Kantor, P. 168

Karachi 66, 128
Keck, M. 64, 74
Kenya 76–7, 83, 84, 85–6
Khrushchev, N. 133
King, A. D. 5, 117, 122, 148,
    168, 174
King, K. 5, 135
knowledge
  distributed 19
  experiential 3, 71
  functionalist view of 18
  lay 65
  legal 58, 59, 61, 65, 175, 176
  new-shared 97
  non-technical 71
  organizational 3
  politics of 65
  practical 54
  rational 3
  situated 18
  technical 65, 71, 101
  travelling 16, 17
  see also codified knowledge; local
    knowledge; tacit knowledge; urban
    knowledge; *also under following
    headings prefixed* 'knowledge'
knowledge alliances 65
knowledge creation 4, 72, 76, 12
knowledge production 18, 135,
    168, 172
  academic 171
knowledge transfer 113, 135
  functionalist assumption of learning
    as 63
Knox, P. 38
Kofman, E. 2, 166
Kolkata 42
Kothari, U. 94
Kraftl, P. 21
Kuala Lumpur 139
Kublai Khan 174, 175, 181
Kumbaya-Senkwe, B. M. 136
Kurtilla, T. 49

Lacis, A. 41–2
Lagos 122
Larner, W. 148, 156

Lascoumes, P. 96
Latin America 122
  see also Brazil; Chile; Ecuador; Mexico
Latour, B. 5, 17, 20, 22, 24, 26, 37, 78, 79, 82, 84, 85, 159, 160, 162
Lave, J. 15
Le Corbusier 14
Le Heron, R. 148
Leach, M. 64–5
Leadbeater, C. 5
League of Nations 121
learning
  anthropomorphizing 76
  assembling urbanism through 182
  central claim or assumption about 6
  city cluster 4
  coordinating 19–20
  craft 76
  critical 156, 160–7
  democratization of 167
  entrepreneurial 85–90
  experimental 108
  hidden 39
  ideological 152
  materializing 69–74
  neoliberal 152
  organizational 3, 4–5, 22
  politics of 65, 84, 91
  representation and 74–85
  urban change and 4–6
  see also comparative learning; incremental learning; learning-through-dwelling; tactical learning; urban learning
learning-through-dialogue 102
learning-through-dwelling 21, 37, 43, 47, 49, 53
  help to shape possibilities of 23
  incrementalism central to 9, 38
  modes of 22, 32
  tactics and 54, 55
  unsettling the spatiality of 22
  urban dreaming and 72
  walking a key form of 50
Leavitt, J. 95
Lebas, E. 2, 166

Lefebvre, H. 2, 48, 50, 52, 105, 154, 157, 166, 169
Legg, S. 26, 27, 122, 127, 174
Leith, A. 123, 125, 126
Leitner, H. 68
Leydesdorff, L. 5
Li, T. M. 27, 30, 65, 156
Lingis, A. 21–2
linguistic communism 100
Liverpool 127
local knowledge 40, 136
  external knowledge/ideas superior to 137, 152
Lomnitz, L. 40
London 11, 87, 93, 107, 108, 109, 126, 127, 138
  Charlton 124
  Great Fire (1666) 20
  Whitechapel 124–5
Los Angeles 53
low-income housing/groups 36, 73, 108, 136, 177
  African-American 47
Lumambo, P. 136
Lyons, M. 94

Machlup, F. 5
MacKinnon, D. 4
MacLeod, G. 119, 142
Maharashtra 112, 161
Maharashtra Housing Board 88
  see also MASHAL
Mahathir bin Mohamad 139
Mahila Milan 68, 71, 75, 79, 80–1, 86, 109, 112
Malad 46
Malawi 113
Malaysia 141, 144, 148–9
  see also Kuala Lumpur
Maluf, P. 103
Manchester 124
Manchester School of African Studies 169
Manhattan Institute 12, 117, 134, 137, 138, 152
Mankhurd 68
Marco Polo 174

Marcuse, H. 153, 154, 157
marginalization 28, 41, 54, 57, 59, 74,
    105, 106, 113, 122, 135
  informal settlements 39
  knowledge 64
  multiple 46
  participation and 11, 14, 65, 92, 103
  political control and 93
  progenitor of polarization and 156
  urban capital and 46
  views and practices 125
Maringanti, A. 167
Marsh, D. 118
MASHAL (Maharashtra Social
    Housing and Action League) 71
Masser, Ian 5, 117
materialism
  relational 18
  urban 35
  vital 161, 163
Maxwell, S. 110–11
Mbaye, A. 71
Mbembe, A. 143–4, 151, 169, 171
McCann, E. J. 2, 4, 5, 24, 117, 118,
    134, 145, 148, 149, 150
McDonough, T. 48, 52, 105, 157
McEwan, C. 170, 171
McFarlane, C. 5, 18, 21, 24, 46, 66,
    68, 69, 71, 87, 120, 122, 123, 135,
    164, 168, 169
McGrath, S. 5, 135
McGuirk, P. 24, 30, 144, 147, 156
McKinsey and Company 151
McLuhan, Marshall 128
Memphis 138
Merrifield, A. 52
Mexico 109
Middle East 128
Middleton, J. 49, 50
Mitchell, D. 166
Mitlin, D. 71, 72–3, 83, 87
model house construction 66, 69–74,
    85, 90, 175, 180
  tactical learning in 10, 63, 177
Mohapatra, B. N. 79, 81
Mol, A.-M. 36
monumentalism 12, 116, 131, 133
Morumbi 33

Moscow 131, 133
Moser, C. 40
Mouffe, C. 94
Mozambique 183
MSC (Malaysian Multimedia Super
    Corridor) 141, 149
Mufti, A. 168, 169–70
Mumbai 10, 11, 33, 37, 48, 66, 71, 72,
    81, 134
  discourses of transforming 151
  impoverished settlements 40
  low-income housing 36
  pavement dwellers 74, 76, 78–80, 83
  pilfering of water by slums 160
  sanitation 13, 40, 87, 123, 124,
    126, 150
  street children 9, 43–7, 60, 176
  street hawkers 41, 56
  urban planning 116
  see also Bandra; Byculla; Federation
    of Tenants Association; Rafinagar
Mumbai Slum Dwellers
    Federation 161, 179, 182
  see also NSDF; RSDF; SDI; SRA
Munt, S. R. 51
Muslims 57, 144
Muungano 85–6

Naidu, C. 139
Nair, M. 43
Namibia 83, 113
Naples 42, 48
Nasr, J. 5, 117, 122
Negri, A. 157–8
neighbourhood associations 99, 102
neoliberalism 5, 12, 26, 29, 57, 93,
    103, 116–17, 121, 148, 153, 156,
    161, 182
  circulation of ideology 119
  contemporary policy mobility
    152, 178
  opposition to 67, 105
  short-termist concerns and 94
  see also urban learning assemblages
    (neoliberal)
Neuwirth, R. 39
New Orleans 116, 137–8, 146, 152
New Public Management 106

New York 53, 134, 141–2, 143, 167
  homeless women 74
New Zealand 106, 118, 150
  see also Christchurch
NGOs (non-governmental
  organizations) 44, 47, 58, 62,
  65, 66, 75, 81, 83, 84, 85, 87,
  90, 120
  see also Bombay First; Dialogue on
  Shelter; Groundswell; MASHAL;
  People's Dialogue; SPARC
Nicolini, D. 3, 5, 17
Nicoll, K. 6, 141
Nigeria 112
Nijman, J. 168
Noble, G. 36
Nonaka, I. 3, 5
Norway 87
Nottingham 108
NSDF (National Slum Dwellers
  Federation) 67–8, 75–6, 89,
  109, 112
  see also RSDF
Nuttall, S. 143–4, 171

Obrador-Pons, P. 21
Olds, K. 31, 139
Oliviera, N. D. S. 168
Olssen, M. 141
Ong, Aiwha 24, 26, 29–30, 57, 144,
  145, 147
openness *see* intensity-openness-quality
Orçamento Participativo 99
Otlet, P. 121
Otter, C. 126

Pakistan 128
Pamoja Trust 77
Paraisópolis 33, 49
Paris 48, 125n, 161
Paris Commune 105
Parnet, C. 24
Parr, H. 78
participatory budgeting 14, 94,
  99–106, 111, 154, 178
Patel, S. 67, 71, 72–3, 76, 79, 80–1, 84,
  107–8
Paulick, R. 131

Peattie, L. 95
Peck, J. 4, 5, 28, 115, 117, 119, 121,
  129, 134, 137, 138, 139, 140, 150,
  152, 156
People's Dialogue (NGO) 89
perception 3, 21–3
Perera, N. 122, 127
Philippines 75, 112
Philo, C. 78
Phnom Penh 66, 68, 82
Pieterse, E. 33, 48
Pieterse, J. N. 94
Pile, S. 52, 181
Pinochet, Gen. Augusto 147
planning
  regional 147
  socialist 128
  see also urban planning
Plymouth 127
police 125, 143
  beatings by 46
political repression 66
politics 55, 62–91, 93
  racial 142
  see also ethico-politics
poor people 66, 67, 71, 90, 183
  counting 74–85
Popkewitz, T. S. 141
Porto Alegre 14, 92, 98–106, 107, 113,
  154, 178–9
PowerPoint™ 84, 118, 147, 148
Prabhu, C. 58, 59, 60, 61,
  176, 180
Prakash, G. 46
Prince, R. 145, 150
privatization 65, 136, 138, 141, 143
  service delivery 135
  water 66, 160
progressive taxation 101
Protea South 71
Providence, RI 138
Provoost, M. 128, 129
PT *(Partido dos Trabalhadores)* 62,
  98–9, 102–3, 104
Pune 71, 73
Pyla, P. 130

quality *see* intensity-openness-quality

Rabinow, P. 27, 122
Rafinagar 40, 162
Rajchman, J. 22
Rajdhani Express 44, 45
Ramsamy, E. 134, 136
real estate 88, 141, 156
    escalation of prices 66
    exponential increases in costs 57
reciprocity 40, 41, 42, 60, 112,
    175, 176
    collective recognition and 157
    networks of 140
refugees 44
Rendell, J. 51
revanchism 5, 119, 134, 141, 142, 153
rhythms 4, 8, 23, 33, 47, 49, 51, 181
    diverse 52
    everyday 32, 53
    multiple 9, 48, 53–4, 60, 176, 184
    predictable 27
    spatiotemporal 9, 50, 60, 176
    textual 40
Riles, A. 77, 78, 79
Rio de Janeiro 99
Rio Grande do Sul 106
Risse-Kappen, T. 119
Roberts, J. 20, 22
Robinson, J. 168, 169, 170
Rockefeller Foundation 87, 121
Rose, R. 118
Rossbach, A. 62, 85
Rossi, Aldo 131
Routledge, P. 67
Roy, A. 32, 39, 168
RSDF (Mumbai Railway Slum
    Dwellers Federation) 73

SAHPF (South African Homeless
    People's Federation) 67, 70, 89
Said, E. 111, 170
Salaam Bombay (1988 film) 43, 46
San Diego 140
San Francisco 150
Sandercock, L. 95, 165
sanitation 13, 24, 36, 39, 66, 72,
    122–7, 150
    crisis 12
    groups forced to improvise 40

improved conditions 89
savings groups for 85
women in construction
    programmes 73
Sassen, S. 24, 67
Saunier, P.-Y. 121, 149
Savage, M. 163
savings 75
Savitch, H. V. 168
Schinkel, K. F. 131
Schneider, J. 137
Schumpeter, J. A. 5
scientific modernism 12, 116, 128,
    129, 133, 152
Scotland 143
Scott, A. J. 4
Scott, J. C. 55, 56, 60
SDI (Slum/Shack Dwellers
    International) 10, 13–14, 61,
    62–87, 89, 91, 107, 113, 115, 145,
    148, 163, 171, 175, 180, 182
    meetings in the UK 11
    see also ACHR; Indian Alliance;
    Pamoja Trust; South African
    Alliance
Seattle (protests 1999) 64
seduction 120, 127, 148, 149
Selavip 72
self-development schemes 58, 59
Senegalese Savings and Loan
    Network 71
Sennett, R. 2, 20, 71, 76
Serres, M. 85, 162
Shanghai 134, 151
Shanmughan, L. 75
Sharan, A. 126
Sharp, J. 94
Shove, E. 32
Sikkink, K. 64, 74
Silva, R. T. 103
Simmel, Georg 169
Simon, J. 123
Simone, A. 32, 33, 37, 38–9, 40–1, 48
Simons, M. 141
Sinclair, I. 50–1
Singapore 134, 138
    IT2000 plan 140
    Little India 53

Sintomer, Y. 94, 99, 100, 101, 105, 106
Situationists 157
slum dwellers 56, 66–9
  see also Bangalore Slum Dwellers;
    Mumbai Slum Dwellers; SDI; SRA
Slum Rehabilitation Authority Scheme
    (Mumbai 1995) 88
Slum Sanitation Programme
    (Mumbai) 87
Smith, M. E. 168
Smith, M. P. 29n
Smith, N. 141–2, 168
Smith, S. 21, 36
sociability 27, 49
social movements 19, 54, 57, 61,
    62–91, 98, 129, 157, 182–3, 184
  see also Indian Alliance; Mumbai
    Slum Dwellers; NSDF; RSDF;
    SDI; SRA
socialist realism 12, 116, 130–1,
    133, 148
South Africa 62, 67, 71, 135
  homeless people 108
  poor women 74
  see also Cape Town; Johannesburg;
    People's Dialogue; SAHPF
South African Alliance 69
Southampton 140
Soviet Union 129, 130, 133
  monumentalism 12, 116, 131
  socialist realism 131
SPARC (Society for Promotion of Area
    Resource Centres) 68, 73, 86, 88,
    89, 107, 108, 109
SPARC-Groundswell exchange 112,
    114, 177, 178
Spivak, G. C. 170, 171
SRA (Slum Rehabilitation Authority,
    Mumbai) 58, 88, 90
Sri Lanka 122
  see also Colombo
Stalin, Joseph 130
Star, S. L. 22
Stengers, I. 166
Storper, M. 4
street children 9, 43–7, 60, 142,
    176, 180
Strobel, R. W. 130, 133

strong internationalisation 170–1,
    173, 180
suburbanization 161
Sudan 128
Suplicy, M. 103
surveillance 28, 143
Susser, I. 137
Sutcliffe, A. 5, 117
Swanson, K. 141, 142
Swyngedouw, E. 24, 28
São Paulo 9, 33, 62, 103, 112, 134,
    143, 183
  International Policy Dialogue 112

tacit knowledge 3, 19, 69, 83
  economically valuable 4
tactical learning 9, 33, 54–9, 60–1, 71,
    73, 85, 91, 176, 180
  coordination and 59, 90–1
  examples of 10, 63, 80, 90, 113, 162,
    177, 178–9
  key form of 69
  potential to become spaces of 97
Tampa Bay 139
Tampio, N. 25, 52, 155, 158, 165
Tarakeswar (Kali Temple) 42
tax evasion 101
Thailand 71, 75
Theodore, N. 4, 5, 115, 117, 119, 121,
    152, 156
Theodosis, L. 129, 130
theory cultures 168–9, 170,
    171, 172
  multiple 173, 180
Thomas, A. 121
Thrift, N. 18, 24, 31, 39, 119
Tokyo 148
Tomlinson, R. 135
Transatlantic Summit of Mayors
    (2000) 121
transgovernmental networks 119
translation 9, 16, 23, 25, 30–1, 54, 61,
    76, 78, 79, 81, 82, 117–21, 148,
    169, 176
  chains of 80, 83
  colonial urbanism and 122–8
  distribution, practice and
    comparison 17–19

translation (cont'd)
　how technical knowledge is learnt through 71
　importance of 10, 63
　key form of learning through 150, 151
　multiple domains of 91
　translocalism and 105–13
translocal urban learning assemblages 11, 29, 69, 92, 142, 145, 178, 183
　crucial role of ideology in shaping 134
　emerging 83
　ongoing formation of 102
　relational topology of 120
transport networks 4
travelling policies 24, 115–52
trust networks 140
Turner, J. F. C. 36

Uganda 80
UN Habitat conference (2001) 4, 74, 77, 105, 106, 112, 113, 135, 183
uncertain forums 93–8
UNDP (UN Development Programme) 4, 106
UNESCO (UN Educational, Scientific, and Cultural Organization) 106
Union Internationale des Villes (1913) 121
United Nations 74
　see also UN Habitat; UNDP; UNESCO
United States 87, 118, 121
　Cold War urban policy 128
　see also Austin; New Orleans; New York; San Diego; San Francisco
Upton, R. 117
urban change 1, 4–6, 40, 156, 184
　gradual temporality of 9, 60, 176
　myriad forms of 181
　neoliberal consensus around 143
　radical 90
urban dwelling 32, 69
　acts that remove, deny or radically devastate 40
　continuities and discontinuities of 48

　coordinating device that emerges from experience of 73
　crisis of 91
　everyday 128
　incrementalism a central process of 60, 176
　infrapolitics of knowing that disrupts 55
　liveliness of 48
　unfolding of 53
urban dwelling possibilities 9, 33, 57, 80, 82, 97
　alternative 61, 176
　learning opportunity for rethinking 59
　more equal 154
　new 54, 71, 91, 177
　recast 179
urban forums 1, 20, 180
　dialogic 98–105
　participatory 11, 92, 94, 99, 106
　translocal 111, 114, 183
urban knowledge 20, 41, 153–4, 159, 166, 167, 179
　bourgeois 157
　critical 2
　evaluating existing dominant forms of 13
　production and movement of 18
　travelling 5
urban learning 1, 23, 31, 32, 92–114
　critical geography of 2, 28, 153–73
　locating 6–14
　spatialities of 25
urban learning assemblages 1, 9, 12, 27–8, 30, 31, 91, 116, 128, 129, 133, 175, 177, 179, 181, 184
　attempt to constitute 165
　changing and contingent ideologies that frame 152
　critical geography of 14, 155, 168
　documentary representations within 10, 63
　image as inclusive cosmopolitanism 166
　multiple space-times of intervention within 164
　neoliberal 134–45

see also translocal urban learning assemblages; urban learning forums
urban learning forums 10, 13, 14, 91, 92, 94–104, 108, 114, 115, 154, 166, 178–9, 182
  categories that can militate against prospects of 93
  important for urban policy, planning and activism 105
  new forms of learning that emerge through 175
  translocal 11, 106–7, 109–13, 177, 180, 183
urban planning 19–20, 84, 104, 113, 116, 128–34, 154, 178–9
  clientelistic structures of 101
  colonial 122, 123
  leverage space for the poor in 67
  mobile initiatives 12
  people participation in 65, 95, 103, 106, 183
  translocal 177
Urban Poor Fund International 87
urban space 7, 38, 47, 50, 60, 141, 149, 166, 182
  creatively tinkering with 39
  experiential immersion in 32
  grassroots movements to reshape 48
  how we inhabit or perceive 8, 181
  inventive attempts at appropriation of 157
  repeated, incremental engagement with 40
urbanism 50
  assembling through learning 182
  certain forms have the power to destroy 28
  colonial 5, 117, 122–8, 152, 174, 178
  cyborg 24
  everyday 30, 54, 57
  gentrified and gated 182
  incremental 9, 32, 33–43, 47, 176, 180
  modernist 157
  polyrhythmical nature of 52, 60
  postcolonial 168, 172, 179

progressive cosmopolitan 164, 165–6
rethinking by rendering it strange 51
socially (un)just forms of 20
tactics, politics and 62–91
temporalities and spatialities that shape 53–4
travelling 5, 134
understanding of 6
unitary 157
urbanization 121, 182, 183
  critical elucidation of 28
  lean 136, 137
  rapid 161
USAID (US Agency for International Development) 87
utopianism 64, 148–9
Uttar Pradesh 44, 75

Van Loon, J. 26
Vancouver 148
Vasudevan, A. 144, 162
Venn, C. 24, 53
Verkaaik, O. 32, 46, 47, 54, 55–6, 60–1, 128, 175
Victoria Falls 82, 87
Volait, M. 5, 117, 122

Wacquant, L. 5, 57, 66, 119, 143, 156
walking 52, 54, 55, 79, 80, 86, 171, 176
  conceptual 50, 51
  different forms of 9, 32, 50, 60
  discursive 50, 51
Wang, B. 133
Ward, K. 2, 4, 5, 24, 117, 118, 134, 168
water infrastructures 122
Watson, S. 95
Wenger, E. 4–5, 18
Whitehead, A. N. 159
Wigley, M. 128
Williams, G. 94
Williams, Raymond 90
Williams, Richard 5, 117
Wilson, G. 5
Wolfe, D. A. 4
Workers Party (Brazil) *see* PT

World Bank 5, 12, 74, 87, 94, 105, 106, 117, 134–7, 138, 139, 144, 152, 180, 183
World Conference on Development of Cities (Porto Alegre 2008) 106
World Economic Forum (Davos 1999/2000) 139
World Social Forums 105
Wren, Christopher 20

Wright, G. 122, 125n
Wunderlich, F. 50, 51

Yaeger, P. 169
Yiftachel, O. 46

Zambia 113, 128, 136, 144
Zanu PF 87
Zimbabwe 77, 82–3, 113, 136
  see also Harare; Victoria Falls